COLLECTED POEMS
1987

D. J. ENRIGHT

Oxford   New York
OXFORD UNIVERSITY PRESS
1987

Oxford University Press, Walton Street, Oxford OX2 6DP
Oxford New York Toronto
Delhi Bombay Calcutta Madras Karachi
Petaling Jaya Singapore Hong Kong Tokyo
Nairobi Dar es Salaam Cape Town
Melbourne Auckland
and associated companies in
Beirut Berlin Ibadan Nicosia

Oxford is a trade mark of Oxford University Press

British Library Cataloguing in Publication Data
Enright, D. J.
Collected poems 1987. — Rev and enl. ed.
I. Title
821'.914   PR6009.N6
ISBN 0–19–282061–3

Library of Congress Cataloging in Publication Data
Enright, D. J. (Dennis Joseph), 1920–
Collected poems, 1987.
Rev. and enl. ed. of: Collected poems. 1981.
I. Title
PR6009.N6A17   1987   821'.914   86-31206
ISBN 0–19–282061–3 (pbk.)

Set by Rowland Phototypesetting Ltd.
Printed in Great Britain by
J. W. Arrowsmith Ltd, Bristol

Acknowledgements

This is an enlarged version of *Collected Poems*, 1981, and incorporates poems whose absence from that volume was, if mildly, regretted by readers or by the author. Also added now are a later sequence and a group of recent poems. I am indebted to the editors who first printed many of the inclusions. And likewise to the publishers who first brought most of them out in book form: Routledge & Kegan Paul Ltd for *The Laughing Hyena and Other Poems*, 1953; Secker & Warburg Ltd for *Bread rather than Blossoms*, 1956; Chatto & Windus Ltd for *Some Men are Brothers*, 1960, *Addictions*, 1962, *The Old Adam*, 1965, *Unlawful Assembly*, 1968, *Daughters of Earth*, 1972, *The Terrible Shears*, 1973 (as also Wesleyan University Press, Connecticut), *Sad Ires*, 1975, and *Paradise Illustrated*, 1978; The Library Press, New York/Open Court Publishing Company, Illinois, for *The Typewriter Revolution* (selected poems), 1971; and Oxford University Press, Oxford and New York, for *A Faust Book*, 1979, *Collected Poems*, 1981, and *Instant Chronicles*, 1985.

D.J.E.

Contents

COLLECTED POEMS
1987

THE LAUGHING HYENA
AND OTHER POEMS

Black Country Women

Did they burn in their men's furnaces –
        old women
Whose tired hair coils like cowed and
         dropping smoke,
All shades of grey about their burnt-out skulls?

Small deities of coal, who from your beds
        brought miners –
A narrow street now holds life's remnant,
        and your grey net bags:
You huddle evenings round your fires, each
Boss of a blackened little house.

Perpetual autumn grips this landscape:
        its chopped fields ever dying.
But spring unending flames in its factories:
The Works of men erupt along the countryside.

Swan Village
(Birmingham–Wolverhampton New Road)

Did it turn out a goose? Did some loose liver,
Teetering homewards, run into a sheet?
Or was it some conurban Bernadette, who met
That white display of where the road would never stray,
That dizzy image, never quite forgotten by the busy village?

Or did some fierce and backward god, mad
For a female hand, land with lashing feathers and,
Finding Lemnos but no Leda, leave a
Noun generic to a dismal field, a file of houses, and a derrick?

3

Or, once upon a time, were swans upon a lake
– Like snowflakes on a silver ladle –
Able to evade the falling smuts, the avid poacher?
And is the 'Hen and Chickens' heir
To the singing last sung there?

Or is the bus conductor's cry
No real contention, just a clerical convention?

First Death

It is terrible and wonderful: we wake in the strange night
And there is one bed empty and one room full: tears fall,
The children comfort each other, hugging their knees,
        for what will the future be now, poor things?

And next day there is no school, and meals are disorderly,
Things bought from shops, not the old familiar dishes.
New uncles come from far away, soft-voiced strangers
Drinking extraordinary wines. A kind of abstract kindliness
Fills the house, and a smell of flowers. Impossible to be bad –

Other nights pass, under conceded night-lights and a cloud
Of questions: shall we ever go back to school? Ever again
Go to the pictures? Are we too poor for new shoes?
                                    Must we move
To a council house? Will any of our friends remember us?
Will it always be kind and quiet and sad, like this?

Uncles depart. We go for a week to a country aunt,
Then take a lodger. New shoes are bought – Oh,
        so this is the future!
How long will it last, this time? Never feel safe now.

Waiting for the Bus

She hung away her years, her eyes grew young,
        And filled the dress that filled the shop;
Her figure softened into summer, though wind stung
        And rain would never stop.

4

A dreaming not worn out with knowing,
A moment's absence from the watch, the weather.
    I threw the paper down, that carried no such story,
But roared for what it could not have, perpetual health and liberty
                        and glory.
    It whirled away, a lost bedraggled feather.

Then have we missed the bus? Or are we sure
    Which way the wind is blowing?

An Egyptian in Birmingham

Behind the black barrage the Northern phantoms gather:
Whole winters huddle, weeks of wet Walpurgis, and hour by hour
The Dane's uneasy twilight smothers me, in
Clouds that weep a monstrous European misery –
And my stiff dread of living free and lonely, unconfined and cold.

Although each drop of rain be sibilant, a sibyl, be syllable of history,
Give me my Sun, shameless and gross and full of cruel humours.
Surely the world's burst conscience overhangs this island –
Shall I drown in the sorrows that flood in with the wind?

– Sorrows not my own, when I have more than many?
Starved of my shih-shah, tric-trac, and the click-clack of consoling
                    friends –
How we drowned our lusts in endless anecdote, sweetened
All small humiliations with our honeyed memories!
O Hassan, Ahmed, Ibrahim, our lives had grown to myths –
But now, what fun could even Goha find, in Birmingham?

And if, one dreary morning, I should stand and cry,
My tongue erect with God again, outside the cold clean Civic
                    Centre –
What drama follows, what vast emotions writhe across the busy
                    squares?
A sombre large policeman with a low-pitched voice,
A cautious muffled Justice, snuffling and indistinct.

At home the sea would smile and smile, while near the shore
The suave shark rolled, like some notorious pasha:

We flocked to the sea's edge like a monarch's welcome,
Warmed our hearts with curses, our bellies with the meat of
argument,
Until they sank together, monstrous sun and monstrous fish.

Home is made of love and hate: the beggar's incantation
Clamoured about me where church bells now complain:
Tyranny ran us down in city streets, with klaxons sneering –
Yet Freedom lounged in every alley, no man too poor
To buy her hourly, for a broad story or a smaller coin.
The fat king himself, our daytime devil, we worshipped in dreams,
our hero by night –
In suffering we bought him, weight for weight – he was ours!
Crouched on the smarting sand, my ruined people wept and wailed
Till the great moon came out, royal in blood and gold,
And stared between our heads across a vast and jewelled sky,
Till the tears became laughing, our bruises voluptuous,
And giggling we counted our scars.

But here the crowds stride on like zombies: green and red
The dead lights guide my safe and sullen steps – arms useless
At my sides, my tongue is frozen, its flowers all dead –
But wait, when I go home, you too, Birmingham, you
Shall be marvellous, your fame ripple through the bazaars like a
belly dancer –
A city of peril, like the shark, a city of witty Gohas,
Your streets astream with passions, big as buses,
Brown jinn shall squat upon your chimneys, ravage the countryside –
Your chanting pigeons, at evening, be princesses and enchanted.

When I am home, and my bored days flame up in brilliant story.

Deir El Bahari: Temple of Hatshepsut

How did she come here, when it was new and sparkling,
White and immaculate against the huge and spongy cliffs,
The great Queen, how did she come, in the cool of winter,
To review her voyages, perhaps, or admire her politics?
Not like the tourist, with his camera and serious weary step,
Not like the dragoman, with his sideways twist, poised in a
revelatory date,
Not like the archaeologist, brisk, with a fly-whisk . . .

Grand destination, deserving a green and pleasant air-field,
Or a royal station, garnished with banners and carpets –
How did she arrive, the peaceful Queen, to smile discreetly over her
                              portraits –
Her masculine beard, being man, or her marvellous birth, being
                              god?
To study her exotic wonders, the Red Sea fishes, the fat queen of
                              Punt?
Not like the tourist, homesick on a hell-bred donkey,
Not like the dragoman, informed yet obsequious,
Not like the archaeologist, in a jeep, with new theories . . .

How did she reach here? Kohl-eyed and henna-stained?
Across her breast the whip and the crozier? The desert
Is old and democratic, rude and unpolished the rocks.
The figures of this landscape? During the season,
Gentlemen in shorts and sun-glasses, ladies with tweeds and twisted
                              ankles,
And all the year, the little denizens, with leather feet and tattered
                              gallabiehs,
The wide-striding village women, their drab and dusty dress . . .

But what of a Queen, and one who built this temple, so clean and
                              deep and sure,
Curling inside the clenched and hanging cliff? What magic carpet
Drawn by bright and flying lions? What cloudburst of gold dust?
Between her treasures, incense trees and ivory, panther skins and
                              ebony –
What laid her gently upon those sculptured steps?

The Egyptian Cat

How harsh the change, since those plump halcyon days
                              beneath the chair of Nakht –
Poised in conscious honour above the prostrate fish,
No backstairs bones, no scraped and sorry skeleton,
                              no death's-head
From a toppled dustbin, but sacred unity of fish
                              and flesh and spirit –
With its stately fantail, its fiery markings like your own,
As tiger-god found good its tiger-offering . . .

7

For your seat is now among beggars, and neither man
                              nor cat is longer honoured,
As you with your lank ribs slink, and he
                              sprawls among his stumps.
Young boys, they say, lay their limbs on tram-lines
                              to enter this hard métier –
But you, maddened among the gross cars, mocked by klaxons,
Lie at last in the gutter, merely and clearly dead . . .

And so I think of the old days – you, strong and
                              a little malignant,
Bent like a bow, like a rainbow proud in colour,
Tense on your tail's taut spring, at Thebes –
Where Death reigned over pharaohs, and by dark arts
Cast a light and lasting beauty over life itself – you,
Templed beneath the chair, tearing a fresh and virgin fish.

A Demonstration

Among the starting birds, the gentle hunters stroke their curving
                              boats,
Between the darting fish. Along the moon-washed waters of this
                              splintered lake,
Their beaks peer through the reeds, the thin poles hardly shake its
                              green and living skin.
Here arid unpoetic Africa is suave. No human angers sway them,
As they hunt their birds and fishes. Silent, simple, so absorbed,
These figures, painted in with single strokes, a master's pen.

And then the cry is heard, that their oppression smells to heaven,
That those shall know of it, whose blunted nostrils breathe in
                              Bourse and Palace.
The messenger is blood. And now some unknown native alien –
Far from the Athens that he never knew, the Bourse he never
                              entered –
Drives the fishes into hiding, scares the birds away. Under the
                              sun he smells to heaven.
The Bourse still chokes with old tobacco smoke, the Palace with its
                              royal scents.

University Examinations in Egypt

The air is thick with nerves and smoke: pens tremble in sweating
hands:
Domestic police flit in and out, with smelling salts and aspirin:
And servants, grave-faced but dirty, pace the aisles,
With coffee, Players and Coca-Cola.

Was it like this in my day, at my place? Memory boggles
Between the aggressive fly and curious ant – but did I really
Pause in my painful flight to light a cigarette or swallow drugs?

The nervous eye, patrolling these hot unhappy victims,
Flinches at the symptoms of a year's hard teaching –
'Falstaff indulged in drinking and sexcess', and then,
'Doolittle was a dusty man' and 'Dr Jonson edited the Yellow Book.'

Culture and aspirin: the urgent diploma, the straining brain –
all in the evening fall
To tric-trac in the café, to Hollywood in the picture-house:
Behind, like tourist posters, the glamour of laws and committees,
Wars for freedom, cheap textbooks, national aspirations –

And, further still and very faint, the foreign ghost of happy
Shakespeare,
Keats who really loved things, Akhenaton who adored the Sun,
And Goethe who never thought of Thought.

The Afterlife

Wicked in hand and eye, this knowing arab
Buys, with a stolen stylo, a sweet and scented beverage;
He squats outside the café, and in melting sun
Admires the brown legs of the ladies, and the ocean's frilly edge.

This worthy clerk, whose pen makes others rich,
Shivers in the nameless steam, behind the swaying swinging door;
His hard-earned shilling gains a wet and windy snack.
The rain strikes down, the smoke fights back,
His bus is late once more.

But after life there comes, as this suggests, an afterlife.
The arab shall repent his careless crimes in one long dirty draught
In an eternal midland town. The honest clerk,
Beside a varnished sea, be lulled for ever on a swaying raft,
And chew *soudani*, garnished with dancing girls and waves.

A Winter Scene

(*Hendrick Avercamp, Dutch School*)

The tree uproots herself to join this rout,
About to lift her skirts; the birds look nervous –

Well they might: an easy hundred chubby burghers
Skip and skate and waltz and trip and fall;
And one man seems to drown, but no one cares at all.

Only the boats are frozen, where the elders sit and boast;
Only a disapproving boar looks coldly on – and he will roast.

How brave a colour they make against the sky, against the icy
                              silence what gay din!
Happy ones, the ice they skate on is not thin.

On the Death of a Child

The greatest griefs shall find themselves
                    inside the smallest cage.
It's only then that we can hope to tame
                    their rage,

The monsters we must live with. For
              it will not do
To hiss humanity because one human threw
Us out of heart and home. Or part

At odds with life because one baby failed
              to live.
Indeed, as little as its subject, is
              the wreath we give –

The big words fail to fit. Like giant boxes
Round small bodies. Taking up improper room,
Where so much withering is, and so much bloom.

Among the Falling Leaves

Among the falling leaves, so near the church bells chimed
        the failing hour – we looked to see the tower –
Yet stifled in the autumn's short and rigorous air.
        Funereal lorries paced the sibilant lanes,
Heaved with distaste the dead leaves up in tongs.

It was not holiday, nor holy day: not winter waiting
        on its long reward, nor summer
Thoughtless of the short year's end.
        For in the limbo of the falling leaves
The cafés and the churches all were bare.
Only the public lavatory and telephone still drew
        their silent separate queue.

We walked along the brazen streets, at night, caught
        in a bruised and haggard light,
But met with neither Saxon soul nor Latin spirits.
The lively legions of the poor and maimed
        were long ago reclaimed
By council house and hospital. The latter mystics
Grumbled in Union or plotted in Committee.
        Seduced by television,
The gay rakes baby-sat. And everywhere the sad derision
That kills Today and says, Tomorrow never comes.
        The mind as sharp as knives,
Trapped by his gang of hard facts, cuts his throat.

A meagre moon, between the thin clouds, mopped and mowed,
Recalling her nights of fullness, when she hung
Eager and swollen over molten eastern cities, or softly
Mingled with what the dreamer saw.
        She lit at times our road,
Back to a small house, standing between rumours of war,
Content to stand, mocking the dropping leaves.

Baie des Anges

Was Freud entirely right? We rise to chase those inner phantoms,
Who often end by chasing us. The sleeping dogs
Start up from every corner: they have not read the textbooks
That bid us pat their heads. The only bone they want is us.

11

The villas turn and twist, like orchids, multi-coloured on the
                                    terraced hills.
Amazed, the palm-tree flaunts its plumper head, its feet in fertile
                                    soil;
The olives billow up like smoke, piebald among the darker leaves,
While orange crowds out lemon, the vine creeps when and where it
                                    may,
And roses grow like grass.

And yet the blue sky wanes, the blue sea turns to lead:
The painted villas blanch and shrink, as, massive, metallic,
The clouds climb down the mountains, and dry lightning spurts
                                    among the trees.

Nothing by halves. This richness is a passion that never rests,
From ripeness moves to raging, from the knotted rosebud to the
                                    scattered leaves.
Proud mountain, luminous citron, azure coast – they do not grow
Without the shrivelling thunderbolt, the tired and flaking walls.

The storm passes and the bay once more is full of angels.
The sea's harp ripples and the air is sharp with sudden scents:
Cafés and cars disgorge – good, bad, indifferent – and the bay is full
                                    of men.

Was Goethe wholly wrong? It is by onward striding
We lay our ghosts, he said. Seeking neither to avoid nor meet.
No tree stays small through fear of meeting lightning:
The strawberry finds its ripeness in the straw.
                                    They grow, or rest,
In light or darkness. Doing what they have to do,
And suffering what, and only what, they must.

Death in the South

Whatever snows embrace the soul, you'll have your body lie
In this warm populous land: where its last sigh

Is lost in all the gabble of the frogs, and that high
Sing-song of the tireless crickets. The long cry

Of love and living jars the plump and perfumed sleeping sky
To starry splinters. Oh, remote from that dry

Silence of the stopped machines. And far from where the
                                    nightingale will die,
Sole and eccentric, its histrionic death. You choose to lie

Where life will never let you rest. And if the choirs deny
Their voices, you'll have what the crickets and the frogs supply.

The Laughing Hyena, after Hokusai

For him, it seems, everything was molten. Court-ladies flow in
                                    gentle streams,
Or, gathering lotus, strain sideways from their curving boat,
A donkey prances, or a kite dances in the sky, or soars like sacrificial
                                    smoke.
All is flux: waters fall and leap, and bridges leap and fall.
Even his Tortoise undulates, and his Spring Hat is lively as a pool of
                                    fish.
All he ever saw was sea: a sea of marble splinters –
Long bright fingers claw across his pages, fjords and islands and
                                    shattered trees –

And the Laughing Hyena, cavalier of evil, as volcanic as the rest:
Elegant in a flowered gown, a face like a bomb-burst,
Featured with fangs and built about a rigid laugh,
Ever moving, like a pond's surface where a corpse has sunk.

Between the raised talons of the right hand rests an object –
At rest, like a pale island in a savage sea –   •
                                    a child's head,
Immobile, authentic, torn and bloody –
The point of repose in the picture, the point of movement in us.

Terrible enough, this demon. Yet it is present and perfect,
Firm as its horns, curling among its thick and handsome hair.
I find it an honest visitant, even consoling, after all
Those sententious phantoms, choked with rage and uncertainty,
Who grimace from contemporary pages. It, at least,
Knows exactly why it laughs.

13

Life and Letters

I sat on the parapet, swinging my legs, close under
A luminous sky: a bright night city lay to my right:
Beneath me the seething trams, and a song, long and sad,
From a white café. And history – my own – oh nothing more
    portentous –
Pressed me both ways.

The near stars smelt of jasmine, and the moon – that huge fallafel –
    faintly of garlic.
Electric crickets sang. And bats displayed their talents
In rings around me, which I was too afraid
To fear. It was a time when superstitions drop away.
All day I walked under ladders, forgot to boil the milk.

For history – in the smallest sense – had fallen about me:
Held for a moment between those toppling towers,
Unable to understand, hatefully lost in cheerless ways, I sat
Suspended, dumbfounded, uneasily contained within my debris,
Bare above the hard road, the stiff steel, the tight-faced trams.
Two natives noticed me and jeered: a bored policeman sauntered up:
I went inside.

                              *

Which is why I try to write lucidly, that even I
Can understand it – and mildly, being loath to face the fashionable
    terrors,
Or venture among sinister symbols, under ruin's shadow.
Once having known, at an utter loss, that utter incomprehension
– Unseen, unsmelt, the bold bat, the cloud of jasmine,
Truly out of one's senses – it is unthinkable
To drink horror from ink, to sink into the darkness of words,
Words one has chosen oneself. Poems, at least,
Ought not to be phantoms.

Evening in the Khamsin

Slowly the sea grows pale, and the sun grows more precise,
Shrinking into its slimmer contours, its fear makes us afraid.
The gold in the dying sky dies slowly to silver – but what,
More horribly, is happening in the flattened sea?

Like a frightened cat it crouches, sleek and slightly arched,
The changing lights betray the churning hidden bowels,
A slow sick swell, the tension in the falling line of foam.
What of us, when our blue and boastful sea lies so unnerved?

The silver sun is yellow, now the yellow sun is green: the sky
No longer alight, now the sky is heavy and near. The thick air sighs,
And the light around us slowly moulders – but what,
More dreadful, mercifully distant, happens beneath the sea's tight
    skin?

Not now the enamelled lido, with its brown and holiday bodies,
The confident beach, with the clipped claws of the breaking wave.
But a thick green swirl, it seems, of rotted bodies, putrescent cream –
All the dead of the great seas, gathered together here, still
    inarticulate?

And then in the falling darkness the flying sand is annulled,
Only our cheeks still sting a little, a little grit between the teeth:
The street lamps all burst out, dim yet stronger than that great
    defeated sun,
And across the mollified waters the pathways of gold grow firm.

Autumn Birth

Hard October morning: the telephone said nothing:
Starched sky like linen and the sun far-off and gelid,
Glinting on the stretched road like cutting knives:
Leaves scraped the pavements with small gasps of pain.

From the soft split shell the conker sprang and trembled,
Glistening and new, with its pale and shapeless face.

Soft October afternoon: the telephone sang sweetly:
Sun like a bedjacket fluffed across the loosened sky:
The dead leaves chirped like spring birds in the gutters:
And the long road melted among the sprawling fields.

The weeping conker lay among the gravel: I wiped the dust away,
Tossed it to the crumbling soil: see, it turns over, to sleep, and grow.

BREAD RATHER THAN BLOSSOMS

Sumiyoshi: First Impressions

Last night the lightning whipped the pines,
    and now the sky bears direful signs:
The worst typhoon of all is near.

Well yes, you say, a little early for typhoon,
But soon or late, it is the fate awaited
    every year.

While at my back a gross cicada tunes its
    brassy gear,
A bomb about to burst within my ear.

Next week the earth may quake, and shake
    away the weekend's floods.
I understand why plopping frog in glassy lake
    is plot enough for verse,
And why your songs are tight and brief,
And why a mechanized and ordered death
    affords relief.

Sumiyoshi: After a Typhoon

Pine needles burn along the matted lanes,
The thin astringent smell of this male season stands
In the still light air. Beneath a silver smiling sky
    the bending backs, the busy hands –

Nature affects forgetfulness, but men can not,
    of what has hardly passed –
The still innocuous air, the air of one who has gone
    too far. Above the drooping mast
A softly humming sky disdains to hang its head:
Instead a brisk new race of butterflies averts embarrassment,
    who know of nothing but this pleasant scent
Of broken wood.

Good weather for work outdoors –
      The ladies gather up torn roses,
Tidying their lawn. The women gather up torn houses.
Whatever moral shows itself, these do not notice it –
Landslides absorb them, floods or ravaged roofs preoccupy,
Or ravished waterlilies cry for help.

The Rag-Picker's Feast

    Now he must yield his jovial beard,
The warm red robe and furry boots are taken back.
    He walks away, with pay and frailer perquisites,
Toys bent or broken from the bottom of his sack –

    To strip anew some long denuded neighbourhood;
So quick to know the worn from the worn-out
    And what is trash from what has once been good.
The beard has left an itch, one more to scratch.

    The children do not run to him nor run away;
Unnoticed now he goes, one other wind-swept rag
    That preaches resurrection to the hopeless bins;
As nameless as the cats who share his bag.

    A tinsel strand recalls his cryptic glory –
He wraps it in a paper, reads how foreign Christians cry
    (Their shopping done) 'Put Christ back into Christmas.'
The rag-man hoists his sack, 'Make Christmas longer,' with a sigh.

The Short Life of Kazuo Yamamoto

At the age of thirteen, you passed by the park
Of Nakanoshima, you paused by the Public Library with its
                        well-fed shelves.
The queue outside could have kept you for weeks,
Except that students shine their shoes themselves.

You swallowed the rat poison, all the easier for having a healthy
                        appetite,
And died with admirable definition. Your last words
Were even reported in the papers. 'I wanted to die
Because of a headache.' The policeman took it down, adding that
                        you were quite
Alone and had no personal belongings, other than a headache.

Elsewhere the great ones have their headaches, too,
As they grapple with those notable tongue-twisters
Such as Sovereignty and Subjection.
                        But they were not talking about you,
Kazuo, who found rat poison cheaper than aspirin.

The Pied Piper of Akashi
A Japanese Tale

Despite the striking rows on rows of little stones,
        and large statistics,
Despite the vivid rags and ill-consorting bones –
        a fairy tale alone can make it real and true.

At ten in the morning the black planes flew
        across to bomb the factory
That made black planes. A happy harmless time of day
For children and the aged, both at their various play.

The young ones and old ones scurried to the park,
        the pretty refuge of the useless and the refuse
Of the race. Away from the dark planes in the sky,
        the dark planes on the ground.

But in the morning brightness, the dazed planes found
A human target, by a human error, and let their sleeping
        brothers lie.
They taught the pines a lesson, the grass repented its
        aggression. While nearby
The factory shuddered slightly at the sight.

That night the workers, back from perilous bench or office,
Found their home-town queerly run to middle age –
        no imps of sons, no docile daughters,
And hardly any ancients, whether cracked or sage.

Time, though, and our native riches have once again refuted
    the frozen spell of elves
And witches. New old were soon recruited,
    from ourselves.

The young sprang up afresh, careless of wrongs and rights,
    to shame the frailer rice
And harass our economy. They filled the ownerless kimonos,
    and flew their dusty brothers' kites.

The park is full of bold and bandy babies, and a glory
Of chrysanthemums and paper bags. The nearby factory is full
Of busy adults, glittering planes, and foreign capital –
    a kindly fairy ends our little story.

The Monuments of Hiroshima

The roughly estimated ones, who do not sort well
    with our common phrases,
Who are by no means eating roots of dandelion,
    or pushing up the daisies.

The more or less anonymous, to whom no human idiom
    can apply,
Who neither passed away, or on,
    nor went before, nor vanished on a sigh.

Little of peace for them to rest in, less of them
    to rest in peace:
Dust to dust a swift transition, ashes to ash
    with awful ease.

Their only monument will be of others' casting –
A Tower of Peace, a Hall of Peace, a Bridge of Peace
    – who might have wished for something lasting,
Like a wooden box.

Happy New Year

An avalanche of cards, a thousand calligraphic miles,
    assure a bright New Year.
And yet and yet, I met a banker weeping through his smiles,
    upon that hectic morn.

19

Omedetō, he told me while the saké bubbled on his lips.
His frock-coat trembled in its new and yearly bliss.
'A hard year for Japan,' he said, and gurgled out a sigh.
'A hard grim year.' A saké-smelling tear suffused his reddened eye.

He clapped me by the hand, he led me to his bright new house.
He showed his ancient incense burners, precious treasures, cold and
                                        void.
His family too he showed, drawn up in columns, and his fluttering
                                        spouse.
We bowed and wept together over the grim new year.

I walked away, my head was full of yen, of falling yen.
I saw the others with their empty pockets,
Merry on the old year's dregs, their mouths distilled a warm amen!

The poor are always with us. Only they can find a value in the new.
They are the masters of their fourpenny kites
That soar in the open market of the sky.
Whatever wrongs await, they still preserve some rites.

Tea Ceremony

The garden is not a garden, it is an
        expression of Zen;
The trees are not rooted in earth, then:
        they are rooted in Zen.
And this Tea has nothing to do with thirst:
It says the unsayable. And this bowl
        is no vessel: it is the First
And the Last, it is the Whole.

Beyond the bamboo fence are life-size people,
Rooted in precious little, without benefit of philosophy,
Who grow the rice, who deliver the goods, who
Sometimes bear the unbearable. They too
        drink tea, without much ceremony.

So pour the small beer, Sumichan. And girls,
        permit yourselves a hiccup, the thunder
Of humanity. The helpless alley is held by
        sleeping beggars under

20

Their stirring beards, and the raw fish curls
At the end of the day, and the hot streets cry
    for the careless scavengers.

We too have our precedents. Like those who
    invented this ceremony,
We drink to keep awake. What matter
If we find ourselves beyond the pale,
    the pale bamboo?

Akiko San

She shall be glorified, if any are –
In a modest Jerusalem, planned with an eye for the apt,
She shall roll through the streets in a smart little car
On some angel's expenses, his hands on the wheel.
A wardrobe of jerseys, kimonos and skirts, without
Any patron with a taste to consult. She shall smile
When she wants and be sad if she likes.
                    When she suffers from drought
She shall drink of pure water, not sip watered gin.
Difference of sex none knows in that land of hot springs,
And there with her likes she shall chatter and laugh
                    in the grace-given memory of different things.

Though her camp, for the moment, is beneath an umbrella,
Tendering shelter to the unlikeliest, no matter our station or nation.
Politely resigned to ridiculous offers, she mumbles
The harsh foreign name of her goods, with a blush
                    for her wretched pronunciation.
She tries to look vicious, but merely seems worried,
And pleads with the drunk like a long-lost sister.
But Jerusalem shall glorify her, and take the weight
Off her dumpy legs and soothe the last high-heeled blister.
                    When Death has picked her up cheap,
She shall be left to exercise her natural talents,
                    for mildness and virtue and sleep.

Broken Fingernails

A shabby old man is mixing water with clay.
If that shabby old man had given up hope
(He is probably tired: he has worked all day)
The flimsy house would never have been built.

If the flimsy house had never been built
Six people would shiver in the autumn breath.
If thousands of shabby old men were sorry as you
Millions of people would cough themselves to death.

(In the town the pin-ball parlours sing like cicadas)
Do not take refuge in some far-off foreign allusion
(In the country the cicadas ring like pin-ball parlours)
Simply remark the clay, the water, the straw, and a useful person.

How right they were, the Chinese poets

Only one subject to write about: pity.
Self-pity: the only subject to avoid.
How difficult to observe both conditions!

In mimicked American they invite you to joyless copulation.
Under your feet the begging children threaten your unsteady
                                        balance.
Joyless, you lament the loss of money and of joy.
Fallen down, you bemoan your trousers and your fall.

But make no mistake. Suffering exists, and most of it is not yours.
Good acts are achieved, as good poems are. Most of them not by
                                        you.

The blossoms open their spotless hands, and death falls in.
Death takes us before we become too ridiculous.
White petals do not despise the bleary eye.
A river of rice flows out of a stinking midden.

One thing is certain. However studious we are, or tough,
Thank God we cannot hope to know
The full horror of this world – or whole happiness.

The Old Man Comes to His Senses

Potter's field; where everything is used
And broken;
      in love and strife
The old and new are lost; the ants run
Over all; where art dies back to life.

Mount Hiei scowls; scarred with the sherds
Of history and burnt-out creeds;
      where the bellicose monks
Jostle the mild merchants of *tanka*; and in
The sacred wood, ghostly profundities wail
Among the scaly trunks.
        Where those who have pierced the veil
Are mumbling their thin curds;
The soulful skeletons bat their sleeves.

Drop me in potter's field; where whatever is
      done to me
Is wholly explicable and no more than I deserve –
Among these merry bones; bottles that were drunk
With joy; cups that have overflowed; vases
That held a year of flowers –
      smelling the nostrils of the dog,
Under the flat feet of the poor, knowing the nerve
Of the grasshopper, and the spring of the frog;
      and ants running over all.

A Moment of Happiness

The river-bed is dry. And dry the flesh
Of the long-dead cat. Only the drained fur,
In feathers remaining, spreads a lonely fan.

The sober banners lie in stacks, like fallen leaves, and
Dry the tight lanterns in the phantom shop:
Empty of the light they wait: they thunder
        under the tread of the spider.
The hot iron of the railroad hisses in the air.

It is early autumn,
Waiting between two festivals; the dumb sky thin and blue like
egg-shell.
Till the frugal women, the little aunts bent double
With dry aches, shall suck the heated saké
From their brittle claws.

This bareness pleases me, this hard dry air.
The feast will come. The candle flare
In the paper skull; the skull will grow rosy
And warm; plump fingers ring the cups.
And meanwhile no one talks of national rebirth,
and no one talks of literary renaissances.

A Poor Little Lonely Child Whose Parents Have Gone To A Cultural Festival

Her parents gad about, uneasily spreading culture:
The mother lectures in French, is interpreted into Japanese, and
marks examination papers written in a sort of English.
The father lectures on the impossibility of lecturing on literature,
and having savaged his way through a hundred years of good
literature, takes refuge at home in bad language.

Today is full of a Cultural Festival.
Starting pistols inaugurate it, mysterious banners wave it on.
At 8 a.m. there will be Swedish Drill in the playground.
At 10 a.m. a slightly adapted Nō play will be rendered by members
of the Middle School Society for the Preservation and
Propagation of Traditional National Drama.
There will be dried fish and fizzy lemonade (called 'cyder') available
on the campus from 12 noon onwards.
At 1 p.m. there will be an Exhibition of Flower Arrangement by the
Economics Section in the Gymnasium, simultaneously with an
Exhibition of Baseball by the Society for the Encouragement of
Baseball.
After that you will enjoy two hours of the University Hawaiian
Combo,
followed by two enjoyable hours offered by the Association for
the Enjoyment of Jive.

Culture is a catholic conception. Naturally it includes Geisha.
    But not Street-walkers. Because anybody can walk a street
    whereas not anybody can talk in their noses. You have to draw
    the line somewhere. (And a few of us may have felt a little
    uneasy about letting pro wrestlers in under Culture.)

During the intervals Mammy and Daddy will be attempting not to
    answer unhealthy questions about Sartre and Oscar Wilde and
    Henry Miller. They think people are less likely to get hurt
    playing Baseball or the Electric Guitar. So they lose face by
    recommending George Eliot and Corneille.
Nature too is a part of Culture, and thus they return home cold and
    wet and disgruntled.

In the safety of her four years, she has held a tea-party for her dolls
    all day by the fire.
But now, feeling obliged to add her mite to the Festival,
    she regards for a moment a rather arty plate (Japanese, modern)
    in grey and black,
And remarks: 'That is a good plate. I do not like it, now. Because I
    do not like the colours, now. But that is a good plate.
    And I shall like it when I am bigger.'
With which she returns to her ghastly pink plastic tea-set.

Her parents are dazzled by the plain presentation of a truth that has
    shadowed them all day. They gaze blindly at the plate and hope
    they are bigger.
'An orderly nature,' says the mother, 'A time for this and a time for
    that.'
'Keep her away from visitors,' says the father, 'Praise the Lord, she
    knows too much to be a teacher.'

Blue Umbrellas

    'The thing that makes a blue umbrella with its tail –
    How do you call it?' you ask. Poorly and pale
    Comes my answer. For all I can call it is peacock.

    Now that you go to school, you will learn how we call
        all sorts of things;
    How we mar great works by our mean recital.
    You will learn, for instance, that Head Monster
        is not the gentleman's accepted title;

The blue-tailed eccentrics will be merely peacocks;
      the dead bird will no longer doze
Off till tomorrow's lark, for the letter has killed him.
The dictionary is opening, the gay umbrellas close.

      Oh our mistaken teachers! –
It was not a proper respect for words that we need,
But a decent regard for things, those older creatures
      and more real.
Later you may even resort to writing verse
To prove the dishonesty of names and their black greed –
To confess your ignorance, to expiate your crime,
      seeking a spell to lift a curse.
Or you may, more commodiously, spy on your children,
      busy discoverers,
Without the dubious benefit of rhyme.

Nature Poetry

I was regarding the famous trees, locked in the case
Of a glassy sky, as dignified as some dead face,
As dead as well.
           Until my daughter scampered up, gabbling
Of the famous monkeys in the zoo. And the trees were suddenly
               scrabbling
In the air, they glowed, they shook with communal rage.
For the trees were bereaved of monkeys. And in the zoo
The bitter monkeys shook the dead iron of their cage.

The Individual

      One of nature's errors, a grasshopper
      Livid and long, greener than any grass.
           Yet his vivid song
      Shakes the breathless morning, trusting
      That men will think him a useless flower,
      The animals only notice a precious stone.

Hokusai's Mad Poet

He dances on his naked native toe,
And stars and blots and jottings sport about his head,
Read or unread, his works lie in a silken pile.
Beneath the unpacific sky he dances, while
The autumn pine drops leaves of thought about his head.

One agile line creates him in his twisted robe,
From toe to glorious grin and balding top –
Apply for special status, in return for quip and quirk?
Straightened shall be your crooked line, and stopped your hop!
Pick up your careless leaves and sort your thoughts,
Replace the stars in heaven, modify that smirk!

It is alleged
An empty saké bottle in your company was seen.
No more than drunk you are, on Old Japan –
So far as we can tell, undemocratic Poet San.

Rice

The lightning's slanting eyes pierce through the day.
The pines and palms are petrified. Small torrents rush
Between their stony leaves. The rice plants sway
And trample in their narrow lake. A thousand fish-mouths break the
    slush.
The *kirigirisu* may hop no more. The waterlilies strain
Those eager wrists, their hands are cups for this green raging rain.

– Vast stronghold of what the western books call Nature, and the
    last.
The spring weeds hide the bombing, weeds have hidden Ōmi's old
    imperial city –
This Rice-abounding Land, where crowded hands to crowding
    mouths are passed,
The close closed cycle turns. And in the West
The undecided voices of the grace-abounding lands, in whom we
    rest
Some hopes. In strange accord the sure cicadas sing.
Our bowels grow this rice – but where does that grace spring?

A Kyoto Garden
for Bunshō Jugaku

Here you could pass your holidays,
        trace and retrace the turning ways,
a hundred yards of stepping stones, you feel
        yourself a traveller; alone

you skirt a range of moss, you cross
        and cross again what seem new Rubicons,
a *tan* of land will make ten prefectures,
        a tideless pond a great pacific sea;

each vista, you remark, seems the intended prize –
        like Fuji through a spider's web or else
between a cooper's straining thighs –
        the eyes need never be averted, nor the nose;

across a two-foot fingered canyon, an Amazon of dew,
        a few dwarf maples
lose you in a forest, then you gain a mole-hill's
        panoramic view,

and company enough, a large anthology –
        the golden crow, the myriad leaves,
the seven autumn flowers, the seven herbs of spring,
        the moon sends down a rice-cake and a cassia-tree –

till under a sliding foot a pebble shrieks:
        you hesitate –
what feeds this corpulent moss, whose emptied blood,
        what demon mouths await? –

but then you notice that the pines wear crutches –
        typhoons show no respect for art or craft;
you sigh with happiness, the garden comes alive:
        like us, these princelings feel the draught.

Excusing the Cicadas

Do not take it hard that these verses
    are thick with cicadas –
Those squat persistent creatures,
    one of whom is noise,
Of whom a dozen make a solid silence.
Absorbed as housewives in their larders,
Passionate as tenors, ominous as hearses,
    they tell the whole old story.

I have known them shriek in a cat's mouth
Like a member of the Diet. Or after a *Te Deum*
Tumble abstractedly into the midden. Or,
    gutted by ants,
Pose as elegantly as in a natural history museum.

Lacking politics, philosophy or grace,
Yet they are the blood of many a Japanese poem –
    Here is the monstrous life
Those seventeen syllables cannot embrace.

SOME MEN ARE BROTHERS

Kyoto in Autumn

Precarious hour. Moment of charity and the
            less usual love.

Mild evening. Even taxis now fall mild.
Grey heart, grey city, grey and dusty dove.

Retiring day peers back through paper windows;
            here and there a child
Digs long-lost treasure from between her feet.

Where yesterday the sun's staff beat,
            where winter's claws tomorrow sink,
The silent rag-man picks his comfort now.

The straitened road holds early drunkards
            like a stronger vow;
The season's tang renews the burning tongue.

Poetic weather, nowhere goes unsung,
However short the song. A pipe's smoke prints
Its verses on the hand-made paper of that sky;

And under lanterns leaping like struck flints,
            a potter's novice squats
And finds his colours in the turning air.

A pallid grace invests the gliding cars.
The Kamo keeps its decent way, not opulent nor bare.
The last light waves a fading hand. Now fiercer seasons
            start like neon in the little bars.

Changing the Subject

I had suggested, in exasperation, that he find
  something other to write about
Than the moon, and flowers, and birds, and temples,
  and the bare hills of the once holy city –

People, I proposed, who bravely push their way
  through the leprous lakes of mud.
It was the wet season, rain upon spittle and urine,
  and I had been bravely pushing my way.

It happened my hard words chimed with a new slogan,
  a good idea, since ruined –
'Humanism'. So I helped on a fashion, another like
  mambo, French chanson, and learning Russian.

Now he comes back, my poet, in a different guise:
  the singer of those who sleep in the subway.
'Welcome you are,' his vagrants declaim to each other,
  'a comrade of the common fate.'

'Are they miners from Kyushu?' he asks, these 'hobos
  all in rags.' And adds that
'Broken bamboo baskets, their constant companions, watch
  loyally over their sleeping masters.'

Thus my friend. He asks me if he has passed the test,
  is he truly humanistic,
Will I write another article, about his change of heart?
  I try to think of the subway sleepers.

Who are indescribable. Have no wives or daughters to sell.
  Not the grain of faith that makes a beggar.
Have no words. No thing to express. No 'comrade'.
  Nothing so gratifying as a 'common fate'.

Their broken bamboo baskets are loyal because no one
  would wish to seduce them.
Their ochre skin still burns in its black nest, though a
  hundred changed poets decide to sing them.

'Are they miners from Kyushu?' Neither he nor I will
   ever dare to ask them.
For we know they are not really human, are as apt themes
   for verse as the moon and the bare hills.

The Noodle-Vendor's Flute

In a real city, from a real house,
At midnight by the ticking clocks,
In winter by the crackling roads:
Hearing the noodle-vendor's flute,
Two single fragile falling notes . . .
But what can this small sing-song say,
Under the noise of war?
The flute itself a counterfeit
(Siberian wind can freeze the lips),
Merely a rubber bulb and metal horn
(Hard to ride a cycle, watch for manholes
And late drunks, and play a flute together).
Just squeeze between gloved fingers,
And the note of mild hope sounds:
Release, the indrawn sigh of mild despair . . .
A poignant signal, like the cooee
Of some diffident soul locked out,
Less than appropriate to cooling macaroni.
Two wooden boxes slung across the wheel,
A rider in his middle age, trundling
This gross contraption on a dismal road,
Red eyes and nose and breathless rubber horn.
Yet still the pathos of that double tune
Defies its provenance, and can warm
The bitter night.
Sleepless, we turn and sleep.
Or sickness dwindles to some local limb.
Bought love for one long moment gives itself.
Or there a witch assures a frightened child
She bears no personal grudge.
And I, like other listeners,
See my stupid sadness as a common thing.
And being common,
Therefore something rare indeed.
The puffing vendor, surer than a trumpet,
Tells us we are not alone.

Each night that same frail midnight tune
Squeezed from a bogus flute,
Under the noise of war, after war's noise,
It mourns the fallen, every night,
It celebrates survival –
In real cities, real houses, real time.

Busybody under a Cherry Tree

The pasteboard houses and the plywood schools
Are shaken with their monstrous living load,
Though everywhere blind highways and dark pools
Wait to divert the careless or the indisposed.

Here in a patch of sick maltreated earth,
Much written of but not too proud for this
The cherry comes to its immaculate birth.

This tree reminds the busybody
That falling hopes are not so absolute as falling hairs,
That beauty needs and often gets no civic welcome,
That the half-educated still can love, and theirs
May be the whole which we shall never dare,

And that the cherry's body all year round is busy
Against one week of showered gifts without advice,
For it is silent, for its deeds suffice.

A Pleasant Walk

This sudden wintry country sun
Reminds me of a walk I used to take:
Though that was far away, and in Japan
A road's the only thing they cannot make.

This was an altogether special case,
A noble promenade called Midōsuji:
Contained by banks and trees, it was a place
As sacrosanct though better kept than Fuji.

Pavements ever shining, an air that never stank,
I loitered past the Mercantile, Dai-Ichi Ginkō,
Sanwa, Chase National and Netherlands Handelsbank,
Under the spreading silver of the princely gingkō.

That was a healthy and a pleasing walk:
No beggar cursed the view and no one ever spat
Before the Nippon Kangyō or National City of New York.
High commerce civilizes, there is no doubt of that.

Preserve me from the brutal village sunk in slush;
Allow me Midōsuji, paved in gold from every nation,
Where no dishonoured kitten breaks the busy hush
Of the Hongkong and Shanghai Banking Corporation.

This apparition of a wintry country sun
Reminds me of the walk that now I lack –
Near by the Yodo River, as quiet as any nun,
Where grill and guichet hold the violent city back.

Displaced Person Looks at a Cage-Bird

Every single day, going to where I stay
    (how long?), I pass the canary
In the window. Big bird, all pranked out,
Looming and booming in the window's blank.

Closing a pawky eye, tapping its hairy chest,
    flexing a brawny wing.
Every single day, coming from where I stay
(How long?), I pass this beastly thing.

How I wish it were dead!
    – Florid, complacent, rent-free and over-fed,
Feather-bedded, pensioned, free from wear and tear,
Earth has not anything to show less fair.

I do wish it were dead!
      Then I'd write a better poetry,
On that poor wee bird, its feet in the air,
An innocent victim of something. Just like me.

Sightseeing

Along the long wide temple wall
Extends a large and detailed painting.

A demon's head, its mouth square open,
Inside the mouth a room of people squatting.

Its fangs the polished pillars of the room,
The crimson carpet of the floor its tongue.

Inside this room a painting on the wall,
A demon's head, its mouth square open.

Inside the mouth a room of people squatting,
Their faces blank, the artist did not care.

Inside that room a painting on the wall,
A demon's head, its mouth square open.

Somewhere you are squatting, somewhere there.
Imagination, like the eyes that strain

Against the wall, is happily too weak
To number all the jaws there are to slip.

The Poor Wake Up Quickly

Surprised at night,
The trishaw driver
Slithers from the carriage,
Hurls himself upon the saddle.

With what violence he pedals
Slapbang into the swarming night,
Neon skidding off his cheekbones!
Madly he makes away
In the wrong direction.
I tap his shoulder nervously.
Madly he turns about
Between the taxis and the trams,
Makes away electric-eyed
In another wrong direction.

35

How do I star in that opium dream?
A hulking red-faced ruffian
Who beats him on his bony back,
Cursing in the tongue of demons.
But when we're there
He grumbles mildly over his wage,
Like a sober man,
A man who has had no recent visions.
The poor wake up quickly.

A Polished Performance

Citizens of the polished capital
    Sigh for the towns up country,
And their innocent simplicity.

People in the towns up country
    Applaud the unpolished innocence
Of the distant villages.

Dwellers in the distant villages
    Speak of a simple unspoilt girl,
Living alone, deep in the bush.

Deep in the bush we found her,
    Large and innocent of eye,
Among gentle gibbons and mountain ferns.

Perfect for the part, perfect,
    Except for the dropsy
Which comes from polished rice.

In the capital our film is much admired,
    Its gentle gibbons and mountain ferns,
Unspoilt, unpolished, large and innocent of eye.

Down in the Forest

Silkworms

There was dismay in the famous northern city
As the Buddhists objected to the smothering of silkworms
(Which is done to prevent them biting their way
Out of the cocoon, which injures the silk).

Millions of silkworms stewing in steam-chambers.
Are silkworms not our little brothers?

Yes but our sisters need to dress themselves in silk;
Or need the work of spinning it.

The crisis lasted till a special worm was unearthed
Who leaves a hole in the top of his cocoon
By which he can make his début without a show of force.

This amenable worm is now left unsmothered
To produce millions of large and fruitful butterflies,
And the guiltless silk embraces the limbs of our little sisters.

So, all silkworms are our brothers once again,
And the good news resumes its progress round the world.
Mosquitoes gaze at the tins of Shelltox,
Wondering what moral it holds for them.

Spotted Deer

The white-spotted deer had lots of everything;
But the black-spotted deer had only a little.

So the white-spots sent the black-spots stacks of food,
For which the black-spots were properly grateful.

The white-spots also sent the black-spots knives and forks
To eat the food with, and the black-spots said, 'Thank you'.

Later the white-spots presented pleasant sets of dishes
For the black-spots to eat their pleasant food off.

These were followed by artistic napkins, all embroidered
'To the Black-spots: From their friends the White-spots'.

A professor white-spot came to teach the black-spots
How to sit at a table and eat their nice food nicely.

Of course the black-spots listened with respect,
For White-spot Edible Aid was full of vitamins and savour.

Then the white-spots equipped the black-spots with guided antlers,
To cope with the red-spots whom the smell of food had attracted.

And the black-spots put on the antlers back to front,
Remarking uneasily that 'Under the circumstances' etc.

A doctor white-spot arrived to instruct the black-spots
In the construction of latrines to cope with the increased etc.

Whereupon the black-spots invoked an ancient tradition
Forbidding the digging of holes in the breast of Mother Earth.

A white-spot cultural envoy was gored by black-spot students,
And the red-spots were attracted by the smell of foodlessness.

This coolness continued till low-level changes took place
In both governments and some new traditions were uncovered.

White Elephant

The white elephant
Was handled with great care and labelled 'Fragile'.
They tethered him to a police post, then phoned
For a very large lorry and empty roads.

They mopped away the elephantine sweat
Every few minutes with pocket handkerchiefs,
Lest the pink patches should run, and the whole dissolve
To a merely metaphorical white elephant.

The creature preened himself and desired a mirror.
'Clearly not all elephants are precisely brothers!'
He trumpeted modestly to himself,
This innocent beneficiary of pigmental accident.

The pink patches were in the oddest places,
But one should not look a gift horse in the mouth.
Technically White, he was a dish for a king
And promotion (head of police post) for the finder.

The teacher, halted in the roadside traffic,
Observed the stroking of an elephant with handkerchiefs.
'The humanity of Man!' he trumpeted. 'Who but He
Would care for a poor old animal with skin disease!'

Entrance Visa

Sand and that was all. That was the land.
A second sun blazed up between our feet;
Our hearts were in our shoes, under the sand.

Did you remark the cruel and lovely sea,
                    and how it smiled and left us?
White electric cliffs? Caves blacker than black can be?

No, we remarked our passports in their limp procession.
We tried to seem just like the photograph inside.
We hoped we looked our age, our sex, and our profession.

We were the Descendancy. Hurt but not surprised.
Atoning for our predecessors' every oath and sneer,
We paid in poverty the rich men's debt.
Oh, we had gladly kissed the watching babies,
                    but their tongues were out and wet.

The sand boiled round our shoes;
We needed horny feet, of course, or else a car.
The shed was one big blood-orange; we the bruise.
We praised the customs officer for his fluent speech.
We wanted to declare our loving views. He wouldn't hear.
And so we crept behind our passports,
From rancorous hand to hand, pretending not to care.

Now all that's past.
We find the country has its seasons,
Each street its shadier side. One bar at least
                    allows us credit;
There's no discrimination in the lizard's wink.
We know which laws we can ignore; and all the reasons.
For one of us has written on the country's ways;
We all have read it.

We pause and joke with a policeman.
We see the tourists being done down in the shops,
And see the shops burnt down and looted by the mob,
And see the mob cut down and beaten by policemen.
Perhaps we don't agree with what we see. We understand.
Perhaps we don't fit in, quite. But who does?
We've paid our way.
                    As for our children, though –
Must they pay too, and pay the ancient debt once more?
Or will they seek revenge for our small shames,
                    to turn again the tables of the law?

We pray they'll find the land a little readier for them,
Themselves a little readier for the land.

The Peaceful Island

The name of this place
Is the sound of the surge of the sea on its sands:
The Peaceful Island we call it in English.

Always fighting for peace,
We felt that for once we ought to enjoy it.
We left a maid, two cats, a dog, to hold the fort;
A small unworried boat conveyed us here.

The islanders' speech is soft and slow:
No aids to their dialect have yet been printed;
It sounds like the surge of the sea on powdery sands;
It is empty of oaths.

Now peace defines itself anew for us. How new!
In the yielding curves of the closely wooded shore,
In the pampas grass's hesitant gestures,
In the sea that is ceaselessly giving and taking
                    nothing at all.
All this beauty would take our breath away,
Had we any left.

The man who brings our milk up from the village
Pours it like a libation. On the brow
Of the hill a goatherd greets us with a blessing.
The buying of figs is a ceremony of gods.

But yesterday I threw a stone at a brilliant bird, unthinkingly;
Then thanked God trembling that it missed.
All night the soft sea murmured to the gaping sand,
Like streams of ink across a bottomless blotter.

Dizzy with lack of sleep
Today I throw stones at a row of empty bottles.
One supplies waitresses to the city, another is a flinty industrialist,
One represents an obscure and well-known poet,
Another is expert in traditions and hopes for war.

The noise of breaking glass attracts the peasants
Who gather round, purring benignly in their dialect,
Till I find I am throwing dangerously wide, and go inside –
To wonder what can be happening back in the city,
The buying and selling underneath the arches,
The papers accumulating their puerilities,
The charlatan voices like the surge of an endless sea . . .
So many people, waiting to be told about peace,
So many skulls to drive it into.

Setting up the not-too-broken bottles,
I kill off the evening.
                    From behind a clump of ferns
A goatherd watches, calmly. For he knows
I know that the boat for the city calls in tomorrow.

Am Steinplatz

Benches round a square of grass.
You enter by the stone that asks,
    'Remember those whom Hitlerism killed'.
'Remember those whom Stalinism killed',
Requests the stone by which you leave.

This day, as every other day,
I shuffle through the little park,
    from stone to stone,
From conscience-cancelling stone to stone,
Peering at the fading ribbons on the faded wreaths.

At least the benches bear their load,
Of people reading papers, eating ices,
Watching aeroplanes and flowers,
Sleeping, smoking, counting, cuddling.
    Everything but heed those stony words.
They have forgotten. As they must.
Remember those who live. Yes, they are right.
    They must.

A dog jumps on the bench beside me.
Nice doggie: never killed a single Jew, or Gentile.
Then it jumps on me. Its paws are muddy, muzzle wet.
Gently I push it off. It likes this game of war.

At last a neat stout lady on a nearby bench
Calls tenderly, 'Komm, Liebchen, komm!
Der Herr' – this public-park-frau barks –
    'does not like dogs!'

Shocked papers rustle to the ground;
Ices drip away forgotten; sleepers wake;
The lovers mobilize their distant eyes.
    The air strikes cold.
There's no room for a third stone here.
    I leave.

No Offence

In no country
Are the disposal services more efficient.

Standardized dustbins
Fit precisely into the mouth of a large cylinder
Slung on a six-wheeled chassis.
Even the dustbin lid is raised mechanically
At the very last moment.
You could dispose of a corpse like this
Without giving the least offence.

In no country
Are the public lavatories more immaculately kept.
As neat as new pins, smelling of pine forests,
With a roar like distant Wagner
Your sins are washed away.

In no country
Do the ambulances arrive more promptly.
You are lying on the stretcher
Before the police, the driver, the bystanders and the neighbouring
                                                    shopkeepers
Have finished lecturing you.

In no country
Are the burial facilities more foolproof.
A few pfennigs a week, according to age,
Will procure you a very decent funeral.
You merely sign on the dotted line
And keep your payments regular.

In no country
Are the disposal services more efficient
– I reflect –
As I am sorted out, dressed down, lined up,
Shepherded through the door,
Marshalled across the smooth-faced asphalt,
And fed into the mouth of a large cylinder
Labelled 'Lufthansa'.

Saying No

After so many (in so many places) words,
It came to this one, No.
Epochs of parakeets, of peacocks, of paradisiac birds –
Then one bald owl croaked, No.

And now (in this one place, one time) to celebrate,
One sound will serve.
After the love-laced talk of art, philosophy and fate –
Just, No.

Some virtue here, in this speech-stupefied inane,
To keep it short.
However cumbrous, puffed and stretched the pain –
To say no more than, No.

Virtue (or only decency) it would have been,
But – no.
I dress that death's-head, all too plain, too clean,
With lots of pretty lengths of saying,

No.

The Quagga

By mid-century there were two quaggas left,
And one of the two was male.
The cares of office weighed heavily on him.
When you are the only male of a species,
It is not easy to lead a normal sort of life.

The goats nibbled and belched in casual content;
They charged and skidded up and down their concrete mountain.
One might cut his throat on broken glass,
Another stray too near the tigers.
But they were zealous husbands; and the enclosure was always full,
Its rank air throbbing with ingenuous voices.

The quagga, however, was a man of destiny.
His wife, whom he had met rather late in her life,
Preferred to sleep, or complain of the food and the weather.

For their little garden was less than paradisiac,
With its artificial sun that either scorched or left you cold,
And savants with cameras eternally hanging around,
To perpetuate the only male quagga in the world.

Perhaps that was why he failed to do it himself.
It is all very well for goats and monkeys –
But the last male of a species is subject to peculiar pressures.
If ancient Satan had come slithering in, perhaps . . .
But instead the savants, with cameras and notebooks,
Writing sad stories of the decadence of quaggas.

And then one sultry afternoon he started raising Cain.
This angry young quagga kicked the bars and broke a camera;
He even tried to bite his astonished keeper.
He protested loud and clear against this and that,
Till the other animals became quite embarrassed
For he seemed to be calling them names.

Then he noticed his wife, awake with the noise,
And a curious feeling quivered round his belly.
He was Adam: there was Eve.
Galloping over to her, his head flung back,
He stumbled, and broke a leg, and had to be shot.

Monkey

Once again the Year of the Monkey is here.
I was born in the Year of the Monkey –
Surely a fellow can talk about himself a bit,
       in his own year?

Monkeys are like poets – more than human.
Which is why they do not take us very seriously.
Not to be taken seriously is rather painful.
       To a corner of my cage I retired, mysteriously,
And had sad thoughts. (They may have been deep.)
Big eyes damp with a semi-permanent tear, my thin hands
       held my heavy head from tumbling into sleep.

But even boredom bored me. I hurled myself from bar
To bar, swung on my aching tail, gibbered and grunted
    with all the expected lack of finesse.
If you do a thing at all, do it well. So I hunted
    avidly for the fleas which in fact I do not possess.

Exercise is always good. Daruma lost his legs
    through meditating overlong and deleteriously:
Now they call him a saint, and use him as a paperweight.
'Thank God for Monkey,' thus spake a kindly spectator,
    backing towards the gate,
'Monkey saves us from taking ourselves too seriously!'

Whereupon he straightened his tie, murmured something
    about an important committee . . .
You can have too much exercise. I went back to my niche,
    put on my thinking face, sad and full of pity.
Proudly refused a banana. All night suffered hunger's itch.

But in the next town – who can tell?
    You may like to know that in Chinese
My name (though I cannot write it, being less well-informed than
    some of the flock)
Bears the meaning, 'Monkey comes to Town'.
I think you will find my name on our posters – down
    there, just below the performing pekinese.
Somebody ought to tell the town I'm coming.
    Next year belongs to the Cock.

Brandenburger Tor

Colder than the weather:
Wrecks from which the flesh has withered,
    Everests of rusting iron –
Ours, a world we cannot call our own.

    So memory, who keeps our waking hours,
Evokes a garden, sweating firs,
    High summer in Japan, a sleepy fan
That nods towards a glassy pond,

Where squats a frog, the backward genius
Of a common place. Polonius
    Himself was less compact of truisms.
A beast that cannot breathe except in spasms.

    Then splash! Can't sit about all day.
Time flies, and so does he to catch a fly –
    This frog whom clement memory has raised,
Ugly, soulful, obstinate, bemused.

    But colder than the winter,
This boneyard of extinguished splendour.
    No frog hops home across the icy pile,
No lizard starts, the very ants turn tail.

    Yet someone's there. Old Faithful.
Ugly, stubborn, foolish, soulful,
    The fallen city's final pride. A dog.
Rooting in the rubble, mad after swag.

    Look, east and west, on these unrotting guts
Of what was once a quite majestic Gate.
    A dog is scratching at it. Something's there.
Something to find. Dust whispers. Ruins stir.

Suma

The pines still hold their hostile poses,
                and the sea strikes cold.
The walls of a horde of later houses
                are furred with mould.
How old the new towns are, how very old!

Prince Genji was lucky in his age.
How cheerless a present exile here!
The silence of the common page, or rage
Of chophouse gossip; under too clear a neon
                the eyes soon blear.
Drinking at nights, where can a drunk recline?
                – in this shrinking cage
Not long the Shining One, where even
                the fireflies repine.

The Proper Due

A thin willow hovers here:
Lovely – lovely in spite of
    The thick drain oozing near
Between sick banks, a vein of evil.

    Lovely because of . . .
For acres of willows look like nothing.
    Beauty defines itself against the dirt,
That telling reflection –
    Like health against a hurt –
Deep in the dark infection.

    Hell is easy to foretell –
Mud without the willow,
    Shallow silence with not a single bell,
Still shadow of exhausted monologue.

    Hard to envisage Heaven –
Acres of willows, haloes in eternal floodlights,
    Ten thousand harps that keep in time?
Our best imagination: sights,
    Free and frequent, of the distant slime.

    So one returns to earth, to see again
The willow, and to pay the filth its proper due.
    That in the end – the very end – one vision
Grows from two.

ADDICTIONS

In Memoriam

How clever they are, the Japanese, how clever!
The great department store, Takashimaya, on the
Ginza, near Maruzen Bookshop and British Council –
A sky-scraper swaying with every earth-tremor,
Bowing and scraping, but never falling (how clever!).
On the roof-garden of tall Takashimaya lives an
Elephant. How did he get there, that clever Japanese
Elephant? By lift? By helicopter? (How clever,
Either way.) And this young man who went there to teach
(Uncertificated, but they took him) in Tokyo,
This Englishman with a fine beard and a large and
(It seemed) a healthy body.
                   And he married an orphan,
A Japanese orphan (illegitimate child of
A geisha – Japanese for 'a clever person' – and a
Number of customers), who spoke no English and
He spoke no Japanese. (But how clever they were!)
For a year they were married. She said, half in Japanese,
Half in English, wholly in truth: 'This is the first time
I have known happiness.' (The Japanese are a
Clever people, clever but sad.) 'They call it a
Lottery,' he wrote to me, 'I have made a lucky dip.'
(She was a Japanese orphan, brought up in a convent.)
At the end of that year he started to die.
They flew him to New York, for 2-million volt treatment
('Once a day,' he wrote, 'Enough to make you sick!')
And a number of operations. 'They say there's a
90% chance of a cure,' he wrote, 'Reversing
The odds, I suspect.' Flying back to his orphan,
He was removed from the plane at Honolulu and
Spent four days in a slummy hotel with no money or
Clothes. His passport was not in order. (Dying men
Are not always clever enough in thinking ahead.)
They operated again in Tokyo and again,
He was half a man, then no man, but the cancer

49

Throve on it. 'All I can say is,' he wrote in November,
'Takashimaya will damned well have to find
Another Father Christmas this year, that's all.'
(It was. He died a week later. I was still puzzling
How to reply.)
                              He would have died anywhere.
And he lived his last year in Japan, loved by a
Japanese orphan, teaching her the rudiments of
Happiness, and (without certificate) teaching
Japanese students. In the dungeons of learning, the
Concentration campuses, crammed with ragged uniforms
And consumptive faces, in a land where the literacy
Rate is over 100%, and the magazines
Read each other in the crowded subways. And
He was there (clever of them!), he was there teaching.
Then she went back to her convent, the Japanese
Widow, having known a year's happiness with a
Large blue-eyed red-bearded foreign devil, that's all.
There is a lot of cleverness in the world, but still
Not enough.

Reflections on Foreign Literature

The stories which my friends compose are very sad.
They border on the morbid (which, in the literatures
Of foreign languages, we may licitly enjoy, for they cannot really
Corrupt, any more than we can be expected to discriminate).

(Sometimes I ask myself: Do I live in foreign countries
Because they cannot corrupt me, because I cannot be
Expected to make the unending effort of discrimination?
The exotic: a rest from meaning.)

('The officer shall engage in no activities whatsoever
Of a political nature,' says my contract, 'in the area where he serves.'
And all activity, it seems, is political.)

Anyway, the stories of my friends are very sad.
I am afraid they are largely true, too, discounting the grace-notes of
                              my elegant friends.
At the heart of the ideogram is a suffering man or woman.

50

I remember my friend's friend, a barmaid in Shinjuku, at a literary
                              pub –
Neither snowy-skinned nor sloe-eyed (though far from slow-witted),
Neither forward nor backward, of whom my friend
(A former PEN delegate) said in a whisper:
'Her life-story would make a book. I shall tell you one day . . .'
The day never came. But I can imagine the story.

My friend's friend also made special ties out of leather;
My friend gave me one, as a parting gift, a special memory of his
                              country.
It has an elegant look; but when I wear it, it chafes my skin;
Whispering that nothing is exotic, if you understand, if you stick
                              your neck out for an hour or two;
That only the very worst literature is foreign;
That practically no life at all is.

Berlin Side-Streets

A street of changing shudders. Trees full
Of almost falling tears, doorways as big as
Houses, and blackened office-windows that
Dismay a couple with their public ghosts –
Oh to be strong and certain like the cars!
Great men have pinned their names at corners.
Dead, alas, dead. Can the florist tell us,
The vintner's buoyant labels help us,
The cigar-shop's oriental scenes suggest?
I know the quiet lunch-rooms, the cafés
Where the cakes are good, the waitresses friendly,
Who know me as a well-known stranger.
Set on a bomb-site, this furniture store
Displays each night, discreetly lighted,
The same bedroom suite, in reduced Empire,
Where lovers pause to take some moral thought,
Sweet it may be, or maybe bitter.
Then, the bare flank of a brand-new theatre,
Deliberate art of a defeated power,
In which the great emotions play in safety,
Off the streets, in municipal safety.
Dead, alas, dead. Avoid public places,
Keep out of queues. For such shows, you must

Book in advance, your diary and your heart
Must carry conviction. Otherwise, the streets –
    (to us, breaking some law indifferent
To him, a grim policeman plays the patron)
    – the streets, the side-streets,
Unlisted schools, unmapped combat areas,
Not to be signalized, not bemoaned, not named –
Others will need them yet, their modesty,
Their incurious stones and compassionate trees.
Truly one finds one's level, and then oneself.

Apocalypse

'After the New Apocalypse, very few members were still in
possession of their instruments. Hardly a musician could call a
decent suit his own. Yet, by the early summer of 1945, strains of
sweet music floated on the air again. While the town still reeked of
smoke, charred buildings and the stench of corpses, the Philharmonic Orchestra bestowed the everlasting and imperishable joy
which music never fails to give.'

                (from *The Muses on the Banks of the Spree*,
                            a Berlin tourist brochure)

It soothes the savage doubts.
One Bach outweighs ten Belsens. If 200,000 people
Were remaindered at Hiroshima, the sales of So-and-So's
New novel reached a higher figure in as short a time.
So, imperishable paintings reappeared:
Texts were reprinted:
Public buildings reconstructed:
Human beings reproduced.

After the Newer Apocalypse, very few members
Were still in possession of their instruments
(Very few were still in possession of their members),
And their suits were chiefly indecent.
Yet, while the town still reeked of smoke etc.,
The Philharmonic Trio bestowed, etc.

A civilization vindicated,
A race with three legs still to stand on!
True, the violin was shortly silenced by leukaemia,
And the pianoforte crumbled softly into dust.
But the flute was left. And one is enough.
All, in a sense, goes on. All is in order.

And the ten-tongued mammoth larks,
The forty-foot crickets and the elephantine frogs
Decided that the little chap was harmless,
At least he made no noise, on the banks of whatever river it used
to be.

One day, a reed-warbler stepped on him by accident.
However, all, in a sense, goes on. Still the everlasting and
imperishable joy
Which music never fails to give is being given.

Confessions of an English Opium Smoker

In some sobriety
I offer to recall those images:
Damsel, dome and dulcimer,
Portentous pageants, alien altars,
Foul unimaginable imagined monster,
Façades of fanfares, Lord's Prayer
Tattooed backwards on a Manchu fingernail,
Enigma, or a dread too well aware,
Swirling curtains, almond eyes or smell?

And I regain these images:
Rocked by the modern traffic of the town,
A grubby, badly lighted, stuffy shack –
A hollow in some nobody's family tree,
The undistinguished womb of anybody's
Average mother. And then me,
In all sobriety, slight pain in neck and back,
Expecting that and then a little more,
Right down to bed-rock.
This was no coloured twopenny,
Just a common people's penny sheet –

To read with cool avidity.
(What would you do with dulcimers,
And damsels, and such embarrassments?
Imagined beasts more foul than real monsters?
No man at peace makes poetry.)
Thus I recall, despite myself, the images
That merely were. I offer my sedate respects
To those so sober entertainments,
Suited to our day and ages.

The Burning of the Pipes
Bangkok, 1 July 1959

Who would imagine they were government property? –
Wooden cylinders with collars of silver, coming
From China, brown and shiny with sweat and age.
Inside them were banks of dreams, shiny with
Newness, though doubtless of time-honoured stock.
They were easy to draw on: you pursed your lips
As if to suckle and sucked your breath as if to
Sigh: two skills which most of us have mastered.

The dreams themselves weren't government property.
Rather, the religion of the people. While the state
Took its tithes and the compliance of sleepers.
Now a strong government dispenses with compliance,
A government with rich friends has no need of tithes.

What acrid jinn was it that entered their flesh?
For some, a magic saucer, over green enamelled
Parks and lofty flat-faced city offices, to
Some new Tamerlane in his ticker-tape triumph –
Romantics! They had been reading books.
Others found the one dream left them: dreamless sleep.

As for us, perhaps we had eaten too much to dream,
To need to dream, I mean, or have to sleep.
For us, a moment of thinking our thoughts were viable,
And hope not a hopeless pipe-dream; for us
The gift of forgiveness for the hole in the road,
The dog we ran over on our way to bed.
Wasn't that something? The Chinese invented so much.

A surprise to find they were government property
– Sweat-brown bamboo with dull silver inlay –
As they blaze in thousands on a government bonfire,
In the government park, by government order!
The rice crop is expected to show an increase,
More volunteers for the army, and navy, and
Government service, and a decrease in petty crime.

Not the first time that fire destroys a dream.
Coca-Cola sellers slither through the crowd; bats
Agitate among the rain-trees; flash-bulbs pop.

A holocaust of wooden legs – a miracle constated!
Rubbing his hands, the Marshal steps back from
The smoke, lost in a dream of strong government.
Sad, but they couldn't be beaten into TV sets;
As tourist souvenirs no self-respecting state
Could sponsor them, even at thirty dollars each.

Sticky Afternoon

In the squares the glamorous fountains,
Trembling, droop uncertain heads.

Taps relapse to trickles. (Soap congealing
On my chest, I curse incompetent companies.)

'Kind indulgence' begged in every cinema:
A break for cat-calls and relief of bladders.

High fidelities falter. False notes
Flourish. Highflown tones fall flat.

Out in a suburb, someone's even hotter.
A servant throws the switch: the whole town helps.

A traitor, of some sort: conspired against the state:
Inspired by powers of darkness: gave us quite a shock.

The Hard Core

You can find them anywhere.
In better managed states, you'll have to look:
They're there, unadvertised behind the hoardings,
In casual self-concealing tenements,
Asleep by public fountains.
In badly managed states, they walk the streets
Free citizens, free to beg.
        One is a junkie. Another armless.
One is spoiled by rheumatism (from working
Long hours on bridges or public fountains).
Some were born too weak to keep their strength up.

The commoner suffer merely from consumption,
And dysentery, and child-bearing, and anaemia.
Nothing, it seems, can dissolve this hard core
Of disorder and disease.

      The anarchist moves among them delicately
(Despite his age), with cast-off bread and
Clothing, small change and opium dottles,
Little gifts for his admirers
(As they would be, if they knew, if they'd had
The chance to know). They prove his point,
Without once opening their surly mouths.
They are the only people who count.
      He counts them.
Sometimes wondering, as he washes his hands
At a public fountain: Is he sorry?
Is he an anarchist because he is sorry?
Is he sorry because he's an anarchist?
Grateful, at last, when one of them spits
In his face. Breaking a law.

Brush-Fire

In a city of small pleasures,
     small spoils, small powers,
The wooden shacks are largely burning.
Bodies of small people lie along the shabby streets,
An old palace is smouldering.
Pushing bicycles piled with small bundles,
Families stream away, from north to south,
From south to north.

Who are these who are fighting those,
Fellow countrymen if not fellow men?
Some follow Prince X, some General Y.
Does Prince X lead the nobility, then,
And General Y the military?
Prince X is not particularly noble,
General Y is not so very military.
Some follow the Prince, for name's sake,
Or family's, or because he is there.
Some follow the General, for last month's pay,
Or family's sake, or because he walks in front.

Princes and generals have moderate ambitions.
An air-conditioned palace, a smarter GHQ,
An Armstrong-Siddeley, another little wife.
The families, driven by some curious small ambition,
Stream away, from east to west, from west to east.
It is in their blood, to stream.
They know what is happening. None of them asks why –
They see that foreign tanks are running off
    with native drivers,
Foreign howitzers are manning native gunners –
As they pass by burning houses, on their way
    to burning houses.
Among such small people, the foreign shells
Make ridiculously big noises.

Baroda

The shy were not too shy;
The nationalistic not too nationalistic;
The Good-Old-Timers not incredible;
The poor were not too poor,
Or not too obvious.

A place where one might leave the heart
Ajar for some small emotion
Suited to the saris and the gardens.
The gardens were soft and silky;
The saris were gardens.

A betel chip built palaces in the mouth;
The Coffee Board's coffee (threepence a cup)
Was frankly enjoyable. As we left,
A small round cloud rolled over in the sky,
Enough to make a very decent tear.

I was a gulli-gulli man's chicken

Come to terms with one's environment, you say?
Grass, grit and weather should be my environment:
I find myself a card, any card, in any pack of cards.

My master cannot do more than he can do.
(The greatest of beings suffers his own limitations.)
My master has almost come to terms with his environment.

Which means: Be startling but not shocking.
Be funny but clean. Be efficient but seem kindly.
He walks a tightrope. The tightrope is me.

Passed under summer frocks, I fight with elastic;
Planted in large bosoms, I beat my head on brassières;
Marooned on barren waists, I skate on whalebone.

Do you know what human flesh smells like, as the
Whole of one's environment? The stuffiness of cotton,
The uncertainty of silk, the treachery of nylon?

Do you know that on the first-class deck, P & O,
My anus is lightly sewn up? The greatest of beings
Must please most of the people most of the time.

My master is all for cleanliness. How much
Cleanliness can a small animal like me endure?
How much being lost, do you think? How much handling?

They have invented toys which look like us.
It's little wonder that we living creatures
Should be held as toys. But cheaper, cheaper.

So I have come to terms with my environment.
Which is: hands that grasp me like a wheel or lever,
And an early (but at least organic) disappearance.

Parliament of Cats

The cats caught a Yellow-vented Bulbul.
Snatched from them, for three days it uttered
Its gentle gospel, enthroned above their heads.
Became loved and respected of all the cats.
Then succumbed to internal injuries.
The cats regretted it all profoundly,
They would never forget the wrong they had done.

Later the cats caught a Daurian Starling.
And ate it. For a Daurian Starling is not
A Yellow-vented Bulbul. (Genuflection.)
Its colouring is altogether different.
It walks in a different, quite unnatural fashion.
The case is not the same at all as that of
The Yellow-vented Bulbul. (Genuflection.)

The kittens caught a Yellow-vented Bulbul.
And ate it. What difference, they ask, between
A Yellow-vented Bulbul and that known criminal
The Daurian Starling? Both move through the air
In a quite unnatural fashion. This is not
The Yellow-vented Bulbul of our parents' day,
Who was a Saint of course! (Genuflection.)

Dreaming in the Shanghai Restaurant

I would like to be that elderly Chinese gentleman.
He wears a gold watch with a gold bracelet,
But a shirt without sleeves or tie.
He has good luck moles on his face, but is not
     disfigured with fortune.
His wife resembles him, but is still a handsome woman,
She has never bound her feet or her belly.
Some of the party are his children, it seems,
And some his grandchildren;
No generation appears to intimidate another.
He is interested in people, without wanting to
     convert them or pervert them.
He eats with gusto, but not with lust;
And he drinks, but is not drunk.
He is content with his age, which has always suited him.
When he discusses a dish with the pretty waitress,
It is the dish he discusses, not the waitress.
The table-cloth is not so clean as to show indifference,
Not so dirty as to signify a lack of manners.
He proposes to pay the bill but knows he will not be
     allowed to.
He walks to the door like a man who doesn't fret
     about being respected, since he is;
A daughter or granddaughter opens the door for him,
And he thanks her.

It has been a satisfying evening. Tomorrow
Will be a satisfying morning. In between
    he will sleep satisfactorily.
I guess that for him it is peace in his time.
It would be agreeable to be this Chinese gentleman.

Making Love

Making love –
Love was what they'd made.
In rooms here and there,
In this town and that.
Something they had made –
Children to a childless pair.

Meeting now to unmake love,
To send this invitee
Who'd overstepped the bounds
Back where he belonged
(Though where did he belong?),
Easy come, they said, he'd easy go.

But love was what they'd made;
Acts turned to things.
These children of a childless pair
Had quickly put on weight,
And now stood round:
'Easier to make us than unmake.'

Such towering children they had made,
Not to be mislaid in hotel corridors,
Or shaken off in trains, or sleep.
And thus walled in,
What move could two small mortals make? –
Except make love.

The Abyss

He walks towards the abyss.
Because, he's told, he owes it to himself.
And because he owes it something,
Because he is drawn by depth and darkness.

So many people walking this road!
Most will turn off a little further on.
But a message comes: To the office please,
Letters await your signature.

He likes signing letters. So returns.
But the abyss still calls. Louder
Than the voice of an air-conditioner.
So he retraces that boring stretch of road.

But is called back again: to compose
A reference for a student, one who left
Well before his time. The files record
'An average student', and nothing more.

This is a challenge to be met,
A challenge to the imagination's power.
The abyss will have to wait. Though he hears
Its voice, like a waterfall heard by a child.

So he retraces that boring stretch of road.
Something inside him starts to lick its lips.
He cannot frame an unpainted picture,
But he frames a rough blurb and a fine critique.

And here at last is the end of the road!
– It is the end of the month as well,
There are cheques to be written out.
Can't they wait? He admits they could.

But a crowd presses at the abyss.
Dangling their plumb-lines, fishing
For samples of its sides and bottom.
No use hanging about. He'll come back later.

Campus in Vacation

The children of peons and labourers
Tell tales on the Library steps, or chase
Gently through dim administrative channels.
Students walk hand in hand, unacademic boy and girl,
Between the silent halls. Now is not the season
For the great performances.

Even the palms are idle, leaves propped together
        like harmless drunks.
It is hot; even at night it is hot;
The strollers look at the moon for a moment's coolness,
A cool assurance that everything changes,
And most things stay the same.

But peace and quiet do the State small service.
'Scandalous,' declares Authority, 'that all
This valuable scientific equipment should lie idle
For months on end!' Voice of our scholarly conscience,
        and our salary.

All this equipment, all this valuable equipment.
The servants' children hide behind fat trees,
The strolling couples unlock their hands,
Seeing a professor address himself to a laboratory,
Firm, valuable, modest, head down, brain up.
He unlocks the door, he unlocks the scientific equipment,
Thinking with gentle scorn of the arts teachers
Lying idle in air-conditioned bedrooms, dreaming
        artistic seductions.

What else will he unlock? What new ground open up?
He gazes through me, panting past his window, after
My unbalanced daughter on her birthday bicycle.
I stick my tongue out. What odds,
Whether the couples walk on the campus and look at
The moon or walk on the moon and look at the earth?
Just so long as there's somewhere left to walk,
        to sit, to cycle,
And something left to look at.

Pitchfork Department

It was patent in this ancient city, paradise of
Statuary, that pigeons lacked respect for greatness.
Lucky statesmen, innocent generals and forgiven thinkers,

Their iron breasts befouled, their noble brows
Turned grey, their swords and croziers rusted,
Manuscripts illuminated, padded shanks gone leprous.

Yet the children loved the pigeons, it pleased the
Taxpayers to be used as perches. They walked our streets,
Sometimes were run over, did not disdain our bread.

So the city fathers, as humane as is befitting
In this age of letters and elections, laid out
Drugged fodder: 'Let the sleeping birds be stacked

With care in corporation vehicles, and conveyed to
Some remote and rural district. Let them there be placed
In appropriate locations in their proper postures.'

They slept the weekend through, lost in a dream
Of the Hall of the Thirty Thousand Buddhas, or the day
When every civil servant shall be issued with a public statue.

On Tuesday afternoon, from under their umbrellas,
The city fathers watched the homing pigeons, assiduous,
    unresenting,
Bowels gently stimulated, natural functions unaffected.

News

It was all recounted, quite some time ago.
'There was my wife, clothed in a hundred patches',
'Corpses are piled on the grass, the smell is terrible',
'Blue is the smoke of war, white the bones of man'.

An old story, left to beginners to cut their teeth on,
To a conventional aside in a colour piece,
Or to Charitable Appeals. What we read about
Is another sort of war, more intellectual, more
Our sort of thing. The real war underneath the war.

Over a squabbling little country – the papers tell us –
A plane belonging to some foreign power shoots at
(And misses) the aircraft of another foreign power.
This constitutes an international incident.

The corpses, well and truly hit, on the grass,
The women in patches, the white hungry children –
This is not an international incident; hardly
A national one. Just an old unreadable story.
We are surprised Tu Fu should have dug it up.

'Why isn't your poetry more personal?'

Well, madam, I was never personal.
Never had the chance to be personal.
Was a poor boy who won a scholarship,
And became a case, a crisis, and
(Though the word wasn't known
Round our way) a symbol.
Life started out on the wrong foot,
And so it continued.
At the age of sixteen I fell in love
(She was, I discovered later, precocious mistress
To the Mayor's prodigal son).
I stood speechless under her bedroom –
She lived in a public house kept by her father
Who had to think of his licence –
Was rewarded with a curl of her curtain's lip
And a lot of rain.
Arrived home as a budding symbol,
With a bad cold in the head.
And so it continued.
I won another scholarship. (Symbols
Tend to repeat themselves.)
And uncles had much to say on this score,
Who might have had to keep our family.
The working class are against symbols.
I cultivated my own garden,
And it grew common or garden weeds.
Went out with an apron of wild oats,
And found the countryside already under cultivation.
I could feel symbolism creeping through my joints.
Was it contagious? My mother didn't catch it.
When doing was in order, I did too little;
When it wasn't, I did too much.
I published a poem
(But dictators never read poems)
And a dictator read it.
I began to suspect that symbols had a sense of humour.
If I mentioned art, it appeared
I was reactionary, or else was radical.
(Who else, through a hack lecture, would make *The Times*?)
I was freedom infringed, or freedom misused.
Was invited to pronounce on large topics
(Including the career of Tagore),

But neither shot nor knighted.
I grew too big for my boots, and
Developed a corn.
Even symbols can lose heart.
Now they have given me up as a bad job –
I shall start to be personal any minute, madam.

THE OLD ADAM

In the Catalogue

It was a foreign horror.
A cold and lonely hour,
A place waste and littered,
And this figure standing there.

Like at first a prized
Cherry sapling swathed in straw.
It was no tree. It was enclosed
In a straw cocoon, and

Wore a hood of sacking
Over the might-be head
And the should-be shoulders.
It seemed to be looking.

What did I fear the most?
To ignore and bustle past?
To acknowledge and perhaps
Find out what best was lost?

It didn't accost. I did.
Rattling in my outstretched hand,
I hoped that money would talk,
A language of the land.

Some inner motion stirred the straw.
My stomach turned, I waited
For its – what? – its rustling claw
Or something I could not conceive.

What happened was the worst.
Nothing. Or simply, the straw
Subsided. 'Please, please!'
I begged. But nothing more.

Fear is glad to turn to anger.
I threw the money down and left,
Heedless of any danger,
Aside from vomiting.

From twenty yards I turned
To look. The shape stood still.
Another ten yards, and I strained
My eyes on icy shadows –

The shape was scrabbling for my coins!
I thanked my stomach. Then
Thanked God, who'd left the thing
Enough to make a man.

To Old Cavafy, from a New Country

'Imperfect? Does anything human escape
That sentence? And after all, we get along.'

But now we have fallen on evil times,
Ours is the age of goody-goodiness.

They are planning to kill the old Adam,
Perhaps at this moment the blade is entering.

And when the old Adam has ceased to live,
What part of us but suffers a death?

The body still walks and talks,
The mind performs its mental movements.

There is no lack of younger generation
To meet the nation's needs. Skills shall abound.

They inherit all we have to offer.
Only the dead Adam is not transmissive.

They will spread their narrowness into space,
The yellow moon their whitewashed suburbs.

He died in our generation, the old Adam.
Are our children ours, who did not know him?

We go to a nearby country, for juke-boxes and
Irony. The natives mutter, 'Dirty old tourists!'

We return, and our children wrinkle their noses.
Were we as they wish, few of them would be here!

Too good for us, the evil times we have fallen on.
Our old age shall be spent in disgrace and museums.

Political Meeting

Nothing human is alien to me.
Except knives, and maybe the speeches
Of politicians in flower.

Dampness, of decay and growth,
Arises all round us,
Indigenous mist from earth's two-way flow:

Can you make out which is which?
I could fall down and rot right on the spot,
Equally I could knife this orator

With all the gusto of youthful delinquency.
Except that the knife is alien to me.
One must hold on to some inhibitions –

Though I still feel the haft in my fist.
God help us, he is only talking,
His expression blurred in the general haze.

A knowing man, doing his job,
Quoting nimbly from several literatures,
Joking with his sworn professional enemies.

Laughter coughs through the mist,
Students hoot genially, a child falls
Out of a tree, bulbs and innuendoes crackle,

And solemn pressmen keep the score
(The workers, perhaps, are working).
We all behave in the manner expected of us.

Then am I pedantic, to look for knives
In the hedges and a mist of blood?
We can't make out our friends, we drift off singly.

Prime Minister

Slowly he ticks off their names
On the long list:
All the young political men.

As he was once himself.
He thinks of how he despised the others
        – the apolitical,
            the English-educated,
            the students he called 'white ants
In their ivory tower'.

Not so long ago, in fact,
He coined that happy phrase 'white ants'.
How he despised them, all they cared for
Was lectures, essays and a good degree!

A small thing these days
        – he tells himself –
To be arrested.
Incredulously he remembers
Not once was he arrested, somehow.

Slowly he ticks off the names
On the list to be arrested.
Tonight, isn't it? Yes,
Between 2 and 4 when the blood runs slow.
The young political men,
Full of fire, hot-blooded.
        – For a moment,
He thinks he sees his own name there.
'Red ants,' he hisses,
Thrusting the list at a waiting policeman.

Meeting the Minister for Culture

– At a party.
In that borderland between
Apologetics and apology:
Neither wishing to revoke,
Neither wishing to provoke.
While tacitly agreeing
That the weather,
The latest drainage schemes,
The brand-new Parliament House,
Are hardly worthy themes.

> He talks about United Nations,
> A large and distant subject –
> In the accent of the region
> (The accent which recalls
> So many pretty girls);
> One listens to the accent
> Rather than his news
> (The accent will last longer
> Than this particular voice,
> These particular views).

Then I talk about my students,
A cultural matter.
And he prefers the past
(Before he was a Minister),
And I prefer the present
(Since I became Professor).
I can't deny his past,
He can't deny my present.
The ice we skate on
Is more than thick enough.

> Then, after all,
> He talks of Parliament House,
> Its brand-new architecture,
> A cultural matter.
> Which blends the occidental
> (Walls of glass)
> With (horn-rimmed roofs) the oriental.
> But is curiously ill-lit.
> Already two have walked alas
> Through its walls unwittingly.

The story takes our fancy.
Teller and listener laugh,
Each in the accent of his country.
I start an explication
Of this transparent allegory –
And then remind myself,
It's no tutorial session –
And fetch beer from the fridge.
Such warmth might melt the ice
On which we skate so nicely.

Misgiving at Dusk

In the damp unfocused dusk
Mosquitoes are gathering.

Out of a loudspeaker
Comes loud political speaking.
If I could catch the words
I could not tell the party.
If I could tell the party
I would not know the policy.
If I knew the policy
I could not see the meaning.
If I saw the meaning
I would not guess the outcome.

It is all a vituperative humming.
Night falls abruptly hereabouts.
Shaking with lust, the mosquitoes
Stiffen themselves with bloody possets.
I have become their stews.
Mist-encrusted, flowers of jasmine glimmer
On the grass, stars dismissed from office.

A Liberal Lost

Seeing a lizard
Seize in his jaws
A haphazard moth,

71

With butcher's stance
Bashing its brainpan
Against the wall,

It was ever your rule
To race to the scene,
Usefully or not.

(More often losing
The lizard his meal, not
Saving the moth.)

Now no longer.
Turning away, you say:
'It is the creature's nature,

He needs his rations.'
And in addition
The sight reminds you

Of that dragon
Watching you with jaws open
(Granted, it is his nature,

He needs his rations),
And – the thing that nettles you –
Jeering at your liberal notions.

Elegy in a Country Suburb

To strike that special tone,
Wholly truthful, intimate
And utterly unsparing,

A man communing with himself,
It seems you need to be alone,
Outwardly unhearing –

As might be now,
The streets wiped clean of traffic
By the curfew

(Apart from odd patrolling
Jeeps, which scurry through
This decent district),

The noise of killing
Far away, too distant
To be heard, above this silence

(A young Malay out strolling
– If you insist on instance –
Chopped by a Chinese gang of boys;

A party of Malays
Lopping an old man's Chinese head,
Hot in their need to burn his hut;

The riot squad,
Of some outlandish race,
Guns growing from their shoulders),

Until tomorrow's news,
And subsequent White Papers,
Analysing, blaming, praising,

Too distant to be heard
Above this heavy hush
Of pealing birds and crickets wheezing,

Tones of insects self-communing,
Birds being truthful with themselves,
Intimate and unsparing –

But birds are always chirping,
Insects rattling, always truthful,
Having little call to twist,

Who never entertained large dreams
Or made capacious claims,
Black lists or white lists:

It hardly even seems
The time for self-communing,
Better attend to nature's artists.

Small Hotel

Not *Guest* –
The Chinese, those corrected souls, all know
A guest is never billed, whereas the
Essence of my aspect is, I pay –

But *Occupier*. Good words cost no more.
The Occupier is hereby kindly warned,
It is forbidden strictly by the Law
– In smudged ungainly letters on his door –
Not to introduce into this room
Prostitutes and gambling, and instruments of
Opium Smoking and spitting on the floor.

By Order, all the lot, *The Management*.
The Chinese have immense respect for Order,
They manage anything you name except

To keep their voices down. Outside my door
The Management all night obeys the Law,
Gambles and introduces prostitutes,
And spits upon the floor and kicks around
The instruments of opium smoking.

It is forbidden to the Occupier
To sleep, or introduce into his room
Dreams, or the instruments of restoration.
He finds he has his work cut strictly out
To meet the mandates of the Law and Order.

Coffee, frying garlic and a sudden calm
Imply the onset of a working day.
Kings and queens and jacks have all departed,
Mosquitoes nurse their bloody hangovers.

So large a bill of fare, so small the bill!
A yawning boy bears off my lightweight bag,
Sins of omission make my heavier load.
Insulting gringo. Cultural-imperialist.

Maybe a liberal tip will mollify
The Law, the Order, and the Management? –
With what I leave behind on that hard bed:
Years off my life, a century of rage
And envy.

The Ancient Anthropologist

Let me tell you how it happened. Once
I had my finger on the pulse,
The pulse of a large and noteworthy people.

This pulse was a pile-driver, a pounder
On golden gates and coffin lids, a grinder
Of organs, a kraken, a jetstream of tears.

Believe me, it was swings and Ferris wheels,
Switchbacks and Ghost Trains and Walls
Of Death, dodgems and multifarious booths.

I dug in my heels, hung on by my nails.
Till 'Hands off!' thundered that great pulse,
Though not so great as not to notice me.

Perhaps it misconceived my fingers pressed
Concupiscently? Or set to twist its wrist?
– Activities that fall outside my field.

That was long ago. Today I'm as you find me.
All my articulations flapping freely,
Free from every prejudice, shaking all over.

Doctor Doctor

It's not the easy life you think, this sanity.
Look –
The streets fall down, and blame you
In cracked voices for expatriate indifference.
The lofty trees
(Which you are ever ready to praise in prose or verse)
Look down their noses.
Is the car stopped? Is it the shadows move?
For twenty minutes you have been talking humanistically
To a stone-deaf whore.
Extending your feet to a baby bootblack
You perceive your shoes have been abducted.
Bar-girls will relieve you of the exact amount.
Do not argue. It is the exact amount.

Including the special extras for special customers
On special nights.
You complain to a fat ice-box, you shake
His icy Chinese hand.
The wrong dialect, the wrong tone, the wrong.
(You think it is easy, all this sanity?
Try it. It will send you mad.)
The street is awash with shadows.
White man, do not run over our shadows!
Roll up that map. The streets have changed their addresses.
A small upright clerk will direct you with detailed
And precise inaccuracy.
Monkeys and children stretch out vague talons.
You say you are a professor?
(Catching, through the narrow slits of flesh,
A glimpse of cheongsam.) Let me tell you,
You look very much like a cowboy to me.
On the walls, eastern godlings spew into pewter pots.
You have knocked everybody's drink over:
Everybody is drinking double brandy.
This blue-eyed moderation is cold and hurtful
To sensitive native. You not like native?
For reasons stated in the Book of Tao Tschung Yu,
But not amenable to translation, we
Do not stock that brand of cigarette.
Shadows lash out like fly-swats.
At the edge of the jungle
Crouch the showrooms of the car-dealers,
At the edge of the showrooms
The latest model crouches in the jungle.
Bow to the mad drivers.
Do not complain.
In this world you have no alibi.
You disagreed with history, now history disagrees with you.
'Guilty' – plead this – 'guilty but sane.'
Bar-girls are adoring the grand madmen, the police
Wave them on with quiet but burning pride.
Your nerves are twitching with sanity,
Like an epileptic you throw yourself down on
Clutch and brake and floor.
Look –

Visiting

Mixing briefly with some
Who've lived for months, years, ages,
Deep down in the abyss,
On lower ledges,

He finds them easy,
Understanding, even almost gay.
How can this be?
He feels the ground slipping away
From under his feet.

So likeable, so considerate,
Yes, even almost healthy
(All their suicides unsuccessful).
How on earth can this be ?

Unless they're visitors from above,
From the gayer, easier flatland?
And he the old dweller in abysses
– Apprehensive, prudent, pained –
On one of the upper ledges?

Well, that would explain
The odd resentment they arouse
In him. And the faint ancient pain
Of drawing breath. And his ragged nails.

Poet wondering what he is up to

- A sort of extra hunger,
Less easy to assuage than some
– Or else an extra ear

Listening for a telephone,
Which might or might not ring
In a distant room

– Or else a fear of ghosts
And fear lest ghosts might not appear,
Double superstition, double fear

– To miss and miss and miss,
And then to have, and still to know
That you must miss and miss anew

– It almost sounds like love,
Love in an early stage,
The thing you're talking of

– (but Beauty – no,
Problems of Leisure – no,
Maturity – hardly so)

– And this? Just metaphors
Describing metaphors describing – what?
The eccentric circle of your years.

Postcard from Hamburg

In the Musikhalle
Yogis demonstrate stomach exercises
In sundry poses.
In the Reeperbahn
Ladies demonstrate bust exercises
In several phases.
In the Planten un Blomen
Plants and blooms demonstrate neck exercises
In varying breezes.

Change

Times have changed.
Remember the helplessness
Of the serfs,
The inexplicable tyrannies
Of the lords.

But times have changed.
All is explained to us
In expert detail.
We trail the logic of our lords
Inch by inch.

The serfs devised religions,
And sad and helpful songs.
Sometimes they ran away,
There was somewhere to run to.
Times have changed.

UNLAWFUL ASSEMBLY

The Accents of Brecht

The sour breath of the underfed
Stood in the overloaded roads.
The farmers sold their daughters,
Daughters sold their flesh and blood.

In that country one almost came
To emulate the accents of Brecht.
Plain speaking was in order,
Plain speaking was merely truth.

These days things are looking up,
One sees more cars on better streets,
Although no revolution intervenes,
No heroes. Nothing much to write about.

Simply: 'One another we must love or
Show no profits,' they explained.
If this put paid to the accents of Brecht,
Who am I then to complain?

What became of What-was-his-name?

Roughly once a quarter
I think of M–.
He must have been inside
For about three years now.
And three months longer
Each time I think of him.

Not that I knew him well.
But he used to hang around the place,
Looking for odd scraps of news.
Not a bad newspaperman,

I used to think,
A graduate with some ability
To express himself,
Comparatively well-mannered,
And properly sceptical
About the official hand-outs.
He would ask you what *you* thought.

Others have gone in
And come out
Since then,
Have taken new jobs and
Fathered new babies.
L– and F– are still inside,
But they're hard-core,
Prison is their favourite freedom.
M– wasn't that sort,
He seemed to me a food and drink man.
But he must have done something very bad.
The papers said nothing about it.

One could ask.
But it would draw attention to you,
And to M–.
They say the cases are reviewed
Every twelve months,
And it wouldn't do him any good
For the Special Branch to submit,
'He was asked after by . . .'

Like the police
In that other country:
'What would a gentleman like you
Want with those fellows?'
No use to answer:
'Well, it's only that
I've been teaching them English
For the past two or three years,
And . . .'
No one believes in pure friendship,
Charity went out
When Aid came in.

'Remember, this isn't Oxford or
Cambridge, you know.'
And then they might ask M–:
'Why would a foreigner be
Interested in you?
What favours have you done this
Foreigner?'
M– had better be totally uninteresting.

Is M– married, I wonder?
I can't remember whether I ever knew.
His wife would have news of him,
But she'll have returned to her family,
To her village.
It's not done
To pursue people to their villages.

Funny,
After three years
A new generation hangs around the place,
Hardly one of them has heard of M–.
It makes you feel your age.

But one thing's certain,
M– is alive.
They don't shoot people hereabouts,
They need to review them every year.
M– is a sort of pet, I suppose,
I remember him every three months.
Not that I ever knew him well.

Processional

for William Walsh

Where are they all? –
The Chancellor and the Vice-Chancellor,
The Deputy Vice-Chancellor and the Registrar,
The Bursar, the Deans of the several Faculties,
The Director of Extra-Mural Studies,
The Estate Officer and the Librarian,
The Chairman of the University Council,
The Esquire Bedell and the Public Orator?

For the scaffolding has collapsed, the
Scaffolding of the impending Science Tower
Has collapsed, with the long thin noise of the
Crumbling of a termite-riddled ivory tower.
And underneath are two female labourers,
Sought for by their colleagues like buried and
Perishable treasure. Now a trousered leg is
Uncovered, and pulled upon, and at its end is an
Ivory visage, a whitened stage concubine's,
Slashed with a vulgar wash of red.

And where is the University Health Physician?
(He is sick, he has left, he is on sick leave.)

And first will arrive the Fire Brigade,
With their hoses and helmets and hatchets
To exhume the already exhumed. And then
The police car with its mild unworried policemen
And hypnotic radio. And last of all,
In accordance with protocol, the ambulance,
To remove whatever the firemen and the policemen
Are no longer interested in.

But where are they all? –
The Development and the Public Relations Officers
And the various Assistant Registrars,
The Vice-Dean and the Sub-Dean of Law,
The Chairman of the Senior Common Room Committee,
The Acting Head of the School of Education,
The Readers and the Senior Lecturers
(The Professors we know are all at work),
And the Presidents respectively of the Local and
The Expatriate Academic Staff Associations?
This is a bad day for an accident.

Till a clerk arrives, a clerk from the
Administration, to administer the matter.
And order is imposed and sense is made.
The scaffolding consisted of old wood left out
Too long in the monsoon rains, and the women
Took too much sand up with them, because the
Contractor told them to get a move on, since he
Was hurrying to finish the job, because . . .
And they fell through four floors,

Carrying the scaffolding with them at each floor,
The sodden planking and the bamboo poles,
And now the scaffolding and the sand and the
Labourers all lie scattered on the ground.
The day and the hour are determined, and the
Victims identified as One Science Tower
(Uncompleted) and Two Labourers, Female, Chinese,
Aged about 20 and 40 respectively,
Who also look rather incomplete.

The Chancellor and the Vice-Chancellor,
The Deputy Vice-Chancellor and the Registrar . . .
There was little occasion for them after all.
The accident has been thoroughly administered –
Moved and seconded, carried and minuted.
A gaggle of idle Assistant Lecturers tap
On their watches, seditiously timing
The ambulance. And in the distance
The fire engine's bell can be heard already.

Shabby Imperial Dreams

I confess with shame
That I have had such dreams.
My only plea is,
I was sleeping at the time –

The winding files of the military,
Who seem to be wearing Roman dress
(Not like this shapeless jungle army
In its carefully tailored rags),

The satraps and their curious recreations,
The consuls and proconsuls
(Not like these gentry in sober suiting
And shrewd but respectful expressions),

And myself, I admit,
A sort of Virgil in these proceedings,
Borne on a litter,
Dictating a letter,
Fed on the choicest local titbits
And the finest imported wines
– Spoilt, just a little,

With a slave girl,
Freed of course,
Set free by me,
To massage me
(Not like the thumbs and fingers
Of our noble Roman matrons),
And amicable natives, gaping gratefully
At an occasional hero hobbling past
(But no foot-rot or leech bites
Or the nausea contracted in the Naafi).

Who would not wish to be lightly looted
By such? What unnatural parent
Could grudge his daughter their attentions?
They speak in measured Latin,
With an English accent.
One of them looks like Enobarbus,
But more mature.
What stories he will have of the frontier!

– I wake to the morning paper,
And stories of expatriate officers
Discharged, sent home with compensation.
Another dream begins,
The dream of self-determination,
Self-empire.

One man's dream is soon
Another's nightmare.
Mine at least are posthumous,
Ruined castles in a vanished Spain

Rueful Writers Contemplate a Painter

The Master, within a week,
Knocks off three paintings,
Well up to standard.

In his tiny studio,
Petrol and pandemonium outside,
And a minimum of view,

Irrespective of situations,
National pride and income,
This bloc and that.

How different from us!
Our stuff the stuff of politicians,
The cries of maddened cabbies

And spavined academics.
The whole day goes on washing words,
Scraping and scrubbing.

Within a week, the Master
Touches sweet oils to virgin canvas,
And makes three paintings.

To be hung,
To be chatted up,
And maybe bought.

Perhaps he even serves the State,
Even that perhaps –
Who can prove the contrary?

At least he does it no disservice,
As we contrive to do,
And sometimes do ourselves.

How shall we serve this Master –
Write notes for his new catalogue?
Old phrases waving grimy fists.

Come to Sunny S

You step down over the silvery sea,
Under the blue air, and there you find
The roads are rather Roman, and Guinness
Is served ice-cold in the wine-shops,
That the buildings are modern, or else
Middle-aged, no older than you are.

The British have left GO-CAUTION-STOP
At intersections. Park opposite a
White line, you will surely be fined.

No mystery, simply a white line, a fine.
The laws are rather Roman too.

There are no nautch-girls here,
No geisha, no sing-song girls, indeed
We have nothing along those lines except
Girls.

Our temples were erected yesterday,
And renovated this morning. We see no
Virtue in decay. Our schools face the old
Problem: children. Our scent is chiefly
Petrol. Our shops sell what we need.

Among our customs, self-sacrifice to
Tourists is not one. We sell you
What you can get at Harrods or Liberty's
But, being the East, a little cheaper.

We do not sell virgins cheap,
We need them. If you prick us, we shall
Probably bleed. For a price you can get
What you can get at home for a price.
We are not certain what we are, we are
Various. You will write books later and
Tell us. We shall display your books
In our tourist hotels.

Our speciality is race riots: which
Wouldn't interest you half as much as
The forbidden dances of temple prostitutes
(Tickets sold at all Government offices)
Or the Spring Festival (held every Saturday)
Of a special class of female artist –
Neither of which we have in this country.

Our speciality is human trouble:
Which mightn't appeal to you, seeing
You could get your heads, breasts or
Balls sliced off accidentally. This
Is our only form of human sacrifice.

And I tell you this because I think
You may be human.

After the Riots

The crickets are making throat-slitting noises.
I lie in the large darkness and cannot sleep.

Jumping with a fear which I am pleased to note
Is not entirely selfish, and sadness more than fear.

And true it is we live in tremulous times,
And the absence of a national consciousness.

Lying and thinking of friends, of a few friends,
Whose throats (who knows?) are being slit this moment.

We each bear all the rest some grudge or other,
Race, religion, colour, food, or even person.

So many grounds for so much throat-slitting,
For the noise of crickets, the noise of crickets.

The Minister desires us all to inter-marry,
Miscegenation is his measured policy.

The crickets are making kiss-kissing noises.
Vast checkerboard of wedlock, black on white.

Alack for Government! – Reduced to venery.
A national consciousness is born in bed.

The crickets sound like sawing wood for coffins.
They sound like someone sawing wood for bedsteads.

Love and death, such grand and soporific themes . . .
The crickets sound like someone making snoring noises.

Unlawful Assembly

This vale of teargas,
More a hospital than an inn.

Clarity begins at home,
How far does it spread?

A gathering is a mob,
Mobs are to be dispersed

Back to their homes
(Lucky to have one)

Back to their jobs
(Lucky to have one)

Why subscribe to clarity?
In this vale of teargas

Should one enter a caveat,
Or a monastery?

Brief Briefing

Colourful? It's green here,
Green and green. And then
A spectrum of ideologies –

Where too much colour
(Discernible red, let's say)
Spells detention.

And too much green
Means jungle here again.
They cut back both.

The spectrum fades to grey,
And grey apartment houses
Elbow out the jungle.

But then the flags of washing
Streaming out from poles –
It's all the colours here!

The Mysterious Incident at the Admiral's Party

Moored in his favourite Eastern port,
This jolly British Admiral
Must give a party on his ship,
With jolly guests, Malays, Chinese,
And Indians and English too.
Says he, 'I like the sarong well,
Trim gear, I wear it when I can.'
Good Jack, who likes Malay costume!
Approval flutters like the gulls,
Down go the drinks, up come the words.
'Although,' he says, 'it tends to slip,
It slips and slides below my hips,
It's hard to keep a sarong up.'
A Chinese lady speaks, sedate
And sweet. 'Then Admiral you need
A songkok.' Songkok as you know
Is headwear proper to sarong.
But Admiral and nearby guests,
They do not know what songkok is,
They think they hear some other words.
Some gape, some giggle and some gasp,
And jolly shaken Jack withdraws
Upon his bridge, and all disperse.
This Chinese lady at a loss,
She asks her spouse in Mandarin,
But what, but why? Who, unbeknown,
Now scouts about the huddled groups.
Then joins his wife. 'Ah me, my dear,'
He murmurs in their tongue, 'To keep
His sarong up – they thought you said –
The Admiral needs a strong –'
'For shame!' in spotless Mandarin
This well-bred lady cries, 'Oh filthy-
Minded foreign hounds! Oh deep disgrace!
What can they think of Chinese dames,
These British gentlemen? Away!'
So Admiral is hurt, Malays
Offended, English persons shocked,
And Chinese lady hates the lot.
Weigh anchor, jolly Admiral –
Let drop these oriental tricks,
Be stayed with buttons and gold braid!

Writing Poetry in a Hotel Room

Is it to tame the place perhaps?
– though heaven knows it's hardly
wild. To beautify the spot?
– it's not exactly ugly.

To clear a space where one can walk
   on homely naked feet?
Or make room in the room
   for possibilities of sleep?

Not home (the waiting word) since
   home is where you left it.
Why organize one here, and
   waste the airline ticket?

Away's away. It's simply, if
   the bedside rug is grubby,
kick it underneath the bed.
   Compound with that nude lady'

who bears a faded lampshade
   on her brassy head. Learn
the quiddities of cupboards,
   where curtains draw the line.

Sow the paperbasket's wastes
   and harvest the *belle vue*.
To open neutral ground
   is one thing poetry can do –

and grow on it: as one might
   even say, a sort of culture. All
the laws of science couldn't white-
   wash that nude lady of the deadly shade;

nor measurement of stress and strain
   lay me soundly down to sleep
in this typhonic wind and rain.
   I'll rather write of sampans

overturning. No, there's no need to –
   I've formed relations with this room,
cool but adequate. I don't mean
   to marry it, and take it home.

Cats and Dogs

One has lived too long perhaps
In a country without cats,
Where the dogs strut by the fence
To keep far hence
The poor man, madman and the thief.

One has lived too long perhaps
In a land of cold-eyed cats,
Where no dogs go
With teeth and tail and tongue to show
Our woes are not unique.

Cultural Freedom

Set free
From all committee,
What would you write?

One writes despite,
In spite of failure,
Of failing light.

One works because
Of lack of leisure;
Out of loss

Of liberty;
To fill deficiency
With presentness.

You need defeat's sour
Fuel for poetry.
Its motive power
Is powerlessness.

Poor Yusoff

Propped against the stand-pipe,
Yusoff contemplates his only friend,
A poem.
Standing there outside his race
(Which stands outside each other race),
At what remove from life
Does Yusoff stand?
He asks himself, that
Silent man, 'Am I
The saddest of the sad,
Or sad among the happiest?'
Saddled with religion,
Fraught with family,
And hobbled with linguistic cares.
There stands Yusoff,
Abstinence to one side,
Continence to the other,
Indigence in the rear.
Poor Yusoff grumbles
(Thinking other people can)
'And I can't blame on riotous living
Verses uncompleted, crumpled sheets!'
Poor naked Yusoff
Props himself against the stand-pipe
(How should he rejoice
At water at his door?
He's no washerwoman).
His future all arranged,
Arranged in ruins,
Coldly sober there he stands, and
Contemplates his only friend,
A poem.
He must save his friend from disrepair.
Yusoff turns the tap on.
Lo!
Words begin to flow.

Roman Reasons

No wide-eyed innocent, he
Had heard tell of villainy,
Had noticed
That one's loot was one's pay.
Yet he could say
– Seeking no comfort in numbers,
In mutuality's immunity –
'I am alone the villain of the earth.'

Finding good reasons
(Reasons are good),
Finding reasons
For the serial assassinations,
For the quotidian killings
(One's pay was one's loot),
The convenient weddings,
And the other treasons.
But not for what he had done.

Despite the logic, the prior logic,
The good, the worthy reasons,
Ratified by that old manual
He carried in his head,
Those *Hints for Roman Soldiers*,
Those counsels scratched in red –

The reasons, the decent reasons,
Each day stronger,
For his leaving their service
– An ageing general, all heart and no brain,
An ageing native woman, wrinkles and wiles –
Who couldn't be served any longer.

And yet, when it was done,
Expeditiously and quietly done –
An emperor, a Jove, a breaker of hearts,
A queen, a Venus, a taker of hearts.
Larger than life – unfair, unfair!
From time to time men die, it would seem,
For love, and the crocodiles eat them.

Healthy, cheerful, realistic. Yet
Somehow unable to excuse himself,
Those *Hints* less lucid than he'd thought,
And logic a botched breastplate.
Alone the villain of the earth. Alone.
So, heart failure in a ditch.
Expeditiously and quietly done,
Unnoticed by *Les Nouvelles d'Egypte*
Or the Roman *Gazette*,
Their columns packed with acts and scenes –

Enobarbus, wine-bearded captain,
You are the hero of my play.

I was a middle-aged corrupter of youth

To wit, a teacher of English literature overseas
And a drunken helot to boot
Maddened with poetry, my hand was half-way up
The nation's sarong. O sinful simplicity!

Why, I would thrust sonnets down the
Throat of a bearded and a turbanned Sikh
I would impregnate the native maidens
With my extraneous versicles

On the buses Malays could be heard
Muttering apophthegms from Milton
Ascetic Chinese boys I would introduce
To some marvellous coy mistress, or Indians
To the alien sophistries of Jane Austen

O yonge fresshe folkes, I would cry
I would impose the dubieties of John Donne
On those who should be in a language laboratory
Learning loyally how to say 'I am a good citizen of
Bahasaland and the kindly British have taught me
How to say so in English'

Who was more affronted than the British Council
To witness the Bahasanese defiled thus?
(Or to see British Literature defiled, I
Could never make out which)

But now I am reformed, I have yielded the promulgation
Of the British way of life to those who understand it
And now I drink my British bitter in a British pub
Surrounded by British elbows and accents
And all is Britishly well with me

But once I was an Overseas Teacher of English Literature
And a drunken helot to boot
– O ars pro arte!

Back

Where is that sought-for place
Which grants a brief release
From locked impossibilities?
Impossible to say,
No signposts point the way.

Its very terrain vague
(What mountainside? What lake?)
It gives the senses nothing,
Nothing to carry back,
No souvenir, no photograph.

Towards its borders no train shrieks
(What meadowland? What creeks?)
And no plane howls towards its heart.
It is yourself you hear
(What parks? What gentle deer?).

Only desperation finds it,
Too desperate to blaze a trail.
It only lives by knowing lack.
The single sign that you were there is,
You know that you are back.

DAUGHTERS OF EARTH

Children Killed in War

A still day here,
Trees standing like a lantern show,
Cicadas, those sparse eaters, at their song,
The eye of silence, lost in soundlessness.

And then, no warning given,
Or if foreseen, then not to be escaped,
A well-aimed wind explodes,
And limbs of trees, which cannot run away,
May only hide behind each other.

Grant their death came promptly there,
Who died too soon,
That pain of parting was not long,
Roots ready to let fall their leaves.

The wind burns out,
The trees, what's left, resume their stand,
The singers stilled, an iron comb
Wrenched roughly through their lives.

While you, your thinking blown off course,
Design some simple windless heaven
Of special treats and toys,
Like picnic snapshots,
Like a magic-lantern show.

Taken Prisoner

On falling into hostile hands
It is advantageous
If you find no tongue in common

So, they are forced to strip their speech
To pidgin, or you
To guess, as is simple, at their few requirements:

Speak out. Shut up. Sit down. Stand up.
Clothes off. Clothes on. Come here. Lie there.

Saved from idiom, from innuendo
From the sounds of childhood, the words of love
The accent of the nerves, the bowels' echo

So, you can say, through the long night, waiting:
These creatures are not like us
They do not speak like us

(Much virtue out of Babel
They howl like werewolves, they grunt like Godzilla
None shall translate them)

So, you are wrapped in silence
In a tower of silence

So, you can say, as the blade is falling:
These creatures are not like us
We are –

So, you forget yourself
You are forgotten

Goodbye Empire

It had to go
So many wounded feelings
And some killings –
In a nutshell, too expensive

Though its going
Scarcely set its subjects free
For freedom –
Life still exacts a fee

In wounded feelings
And also killings
Slates wiped clean
Soon attract new chalkings

At least the old regime
Allowed its odd anomalies
Like my orphaned Irish dad
One of those Wild Geese

Who floundered over India
In the shit and out of it
Getting a stripe, and then
Falling off his horse and losing it.

To an Emergent Politician

One can't make an omelette without breaking eggs, you
Inform us. Which is realistic. Which is even true.

But how many eggs have been broken? And why the haste?
Who is eating the omelette? And how does it taste?

And why do you seem to relish the breaking of eggs more
Than the cooking? The ones lying smashed on the floor –

What are they doing there? May we ask how it happens
You are the only cock left in a huddle of drab hens?

But that's the beauty of imported proverbs for a gent
In your position: the way they choke domestic argument.

An Underdeveloped Country

And there are times truly
(This is no time for irony)
When one can only surmise
That why the whole place isn't bugged
Is the initial cost of the enterprise.

It cannot be because it takes so many
Men to watch so many,
For unemployment's on the up and up
And listening keeps a lot of people quiet.

But one of these days surely
Some big nation
Will stake us to modernization.
We'll end with Russian mikes maybe
And US tape-recorders. We
Are why and where the powers collaborate,
To teach us grown-up ways.

For there are times truly
(Which is no time for irony)
One can't be certain that our friends
And colleagues will denounce us.
They fail to, fairly often.
It seems that we are children
Still, we need that adult help
If we are ever to develop.

J.T. on his Travels

Perhaps you thought she was a child
Perhaps you thought you liked the thought

She doesn't pretend to be a virgin
Her son is lying on the only bed

Almost her size, she carries him
Still sleeping to a next-door cupboard

She brings a bucket of water back
To give herself a local wash

You feel you prefer your own french letter
Hers look secondhand

You she will wash hereafter
If you last that long

Either way you are damned–
If you do it, for unfeelingness

If you don't, for lack of feeling
John Thomas, old Whitey, you can't win.

The Faithful

When I work at nights
I switch the desk-lamp on
And the insects gather.

Small lizards leave their holes
And chase the insects.
In between they squint at me
With nervous admiration,
Tensed to scuttle off:
He gives, He also takes away.

The insects love the light
And are devoured. They suppose
I punish them for something,
My instrument the spring-jawed dragon.

It isn't difficult to be a god.
You hang your lantern out,
Sink yourself in your own concerns
And leave the rest to the faithful.

Board of Selection

No, it is not easy to effect an
Appointment in English literature.
The chairman of the board is worrying over
Last year's riots (the bodies were traced
By the smell . . .) and next year's budget
(The British are pulling out . . . but leaving
Their literature behind them).

101

The dean of the faculty is an
Economist of repute and utility.
But what is the difference, he asks,
Between prose and poetry?
The candidate proposes adroitly that
Poetry is more economical than prose,
Viz., it says as much in half the space.

The economist is not satisfied.
In half the space, he muses . . . but
It takes him four times as long to understand
A piece of poetry as a piece of prose –
Which means . . .

The board make hasty calculations . . .
Which means that poetry is a false economy,
More haste, less speed.
The chairman remembers he has to build a nation
By the end of the month.

Neither I nor the candidate dare ask the
Esteemed economist which particular piece of
Poetry has so discomfited him.
It would probably have to do with daffodils
And this is an orchid-exporting country.
We submit that quite a lot of literature is
Prose and prose is pertinent to the economy.

A business man wants to know when the
Middle Ages stopped and the Renaissance started.
No one is sure. The director of education
Asks why there is so much sex in modern literature.
Because, the candidate ventures, there is so much
In modern life (excluding the English Department).
The chairman is brooding over the birth-rate.

Finally, after a disgruntled pilgrimage to
Canterbury and a brief stopover in 1984,
It is recommended that the candidate be offered
A temporary assistant lectureship at the bottom
Of the scale, subject to the survival of the
Economy, the nation, the university, the department,
And her hopes of completing her master's degree.

More Memories of Underdevelopment

'God's most deep decree
Bitter would have me taste: my taste was me.'

A lapsed Wesleyan, one who dropped out
Halfway through the Wolf Cubs, and later ran howling
From Lourdes by the first bus back, whose idea of
High wit is 'God si Love', who would promptly
Ascribe the sight of Proteus rising from the sea
To spray in the eyes or alcohol in the brain –
Yet these words appal me with recognition,
They grow continuously in terror.

So how much more must they mean
To these young though ageless Catholics, to whom I am
Rashly expounding Father Hopkins!

But no,
They seem to find it a pleasing proposition,
The girls are thinking how sweet they taste, like moon-cake
Or crystallized pears from Peking,
The boys are thinking how good they taste, like crispy noodles
Or bird's-nest soup.

The poor old teacher is muttering curses,
Four letters cut to three out of care for his job:
Was I born to bring a sense of sin among you,
You oriental papists! Obviously Rome was not built in
Three hundred years. A lurching humanist,
Is it for me to instruct you in the fall complete?

Their prudent noses wrinkle almost imperceptibly.
Oh yes, they tell themselves, the poor old man,
His taste is certainly him. . .
And they turn to their nicer thoughts,
Of salted mangoes, pickled plums, and bamboo shoots,
And scarlet chillies, and rice as white as snow.

Master Kung at the Keyboard

for Lee Kum Sing

He's Oriental, he's Japanese, he's Chinese
Watch and you'll see him trip over his tail!
He's a child! What can he know of Vienna woods
Of Ludwig's deafness and J.S.B.'s fine ears?

Of tiaras and galas and programmes
Of hussars and cossacks and pogroms
Of Vespers, Valhallas and Wagrams
And the fine old flower of the Vienna woods?

(Wine, beef, pheasant, cheese, thirst, hunger)

Reared on rice and Taoist riddles
Water torture and the Yellow River
Yang Kwei-fei and one-stringed fiddles –
What can he know of the Water Music
Of barges and gondolas
Of emperors and haemophilias
Of the Abbé, the Princess, and her black cigars?

Wer das Dichten will verstehen
Muss ins Land der Dichtung gehen
Seven days with loaded Canon
Snapping prince and priest and peon.

So he went overseas for his studies? –
It is not in his blood.

What is in his blood?
Blood is.

(Rice, tea, pork, fish, hunger, thirst)

Compared with the minimum of 4,000 characters
Required at the finger-tips for near-literacy
And admission into provincial society
88 keys are child's play.

Play, child!

His heart pumps red rivers through his fingers
His hands chop Bechsteins into splinters
His breath ravishes the leaves
His hair never gets in his eyes.

I am down on my knees.

Every second pianist born is a Chinese
Schubert, Chopin, Mozart, Strauss and Liszt –
He'll be playing on
When the old Vienna woods have gone to chopsticks
Chopsticks every one.

Tourist Promotion

For the tourists, who stay in the
Large new tourist hotels, the
Chief tourist attractions are the
Other large new tourist hotels.

For the querulous and wayward
There were once the local monkeys,
Who lived in the ancient tree-tops
Long before the hotels were thought of.
The tourists enticed the monkeys down
From the trees with monkey nuts and
Breakfast rolls. And the monkeys
Scampered across the road and were
Squashed by the buses transporting
Fresh tourists to see the monkeys.
It was not a pretty sight.

So now the tourists are confined to
The tourist hotels, large and new.
They pass with the greatest of ease
From one to the other, escorted by porters
With large new umbrellas, or even through
Underground passages, air-conditioned and
Adorned with murals by local artists
Conveying impressions of the local scene.
After all, the tourist hotels were created
Specifically for the sake of the tourists.

Valediction

Into your roomy ear I speak,
Your lofty eye will read my lips –

Here's comfort:
Seeing I could never fill you
How should my going leave you empty?

My local lizard
Will hunt across the desk no longer,
Peer from under papers, and
Into the typewriter leap in close pursuit
Of enemy ants –
The desk will not be there.
But he'll be hunting still, a man
Of many dimensions, to whom you offer
Rich terrain.

Peevish bats will squirm in the eaves
To the south; to the north the starlings
Still hold assembly.

The shrew with her questing nose
Shall wheel across the tiles,
Less cover for her then, and thus
The less circuitous her quest. She
Never quested after me.

The swarming bees will piss
Their pollen as before;
Honey was merely meant to be,
Not meant for me.

For cats and dogs
Less of comfort, surely,
But more of interest.
Mosquitoes will miss me most,
The lizards will miss the mosquitoes.

The chain of being is thickly woven,
It does not break if one small link
Drops out. Nature makes do.

The *kempeitai* once came –
And used you as a torture chamber
Or, some oldtimers say, an officers' brothel –
The *kempeitai* then went.
I mention only one
Of your more notable tenancies.

It is not my going
You need to fear, old house,
But the coming of the bulldozer.
The new men have no respect.

Where I am

That cordial cat –
She'd almost purr herself to death.
We were growing old together.

Here, the cry of some new creature
    I can not translate –
Except to guess it grieves!
A sea that breathes of loss alone,
And crickets hissing in outlandish tones
Among outlandish trees.
And then a moon –
There's more than one moon in the sky,
    or else more skies than one!

Lights in the houses here–
    either too soft or hard.
What acts revealed or hidden?
Does love live there? Or hurt?
They eat their evening meal? Or children?

The stranger dies a single death,
Rises to face a hundred lives.

Householders, lend a little light,
Trees and crickets, give me stuff for lines,
Queer moon, permit one rhyme,
    one simple saving rune–
Words, tell me where I am!

Royalties

As 'Name of individual, partnership, or corporation to whom paid'
I find my own, followed by (in brackets) 'Faust'.
The amount of income received in this capacity is $ 3.30 gross,
From which I am glad to see no tax has been withheld.

Egotist as he is
One had never thought the Devil so close-fisted.
He wasn't always:
Gretchen – Helen – contributions to knowledge – all that real
                               estate . . .
What can have changed him?

And yet, Goethe only lasted a couple of months after completing his
                               masterpiece,
So I could even be said to be lucky, nearly twenty-four years
     after publication, still making $ 3.30 out of a little crib
     of the Master's epic designed for non-German-speaking
     near-dropouts taking the World Literature course.

None of us, it seems, even though no tax is deducted, gets much for
                               selling his soul.
The sea took back the land, Gretchen lost her head, Helen was
     incorporeal, the scholarship soon discredited;
Faust died the moment he started to enjoy life; and Goethe's poetry
                               is supplanted by a crib –
It was always a buyer's market, always.

Memoirs of a Book Reviewer

A month much like any other
There were five weekends in it
And no printing strike, but a rape
In Vietnam (I believe it was), a
Symbolical German dentist removing teeth
And illusions, and a Jewish Italian
Afflicted with accidie in an elegant style
(Or maybe that was the month before)
W. Pater wrote letters to an oriental maiden
Raped by GIs, while a bored Italian Jew
Discovered the source of eternal energy in

108

The Antarctic (Prix des Libraires) which was
Discovered a week later by a Swede in a
Documentary novel and a balloon. A number of
Quaint peasants murdered their husbands or
Wives etc. and got off scot-free because they
Were life-enhancing. And I forgot my name
(It was found at the bottom of a page)
I wrote elegant letters to an asian woman
Murdered by the foreign soldiery. Shakespeare
Turned up, with an apt word for everything
Especially titles. I forgot my title
(The postman brought me proofs of existence)
Much drinking of *vins de pays* went on. I can
Remember a hangover. And lots of sex
In a lost world at the bottom of the Antarctic
Discovered by E. Pound, an American balloonist
And I lost my memory for several days, but
Found it at the bottom of a cheque. In a
Spare hour I tried to write a poem on life and
Death in a Vietnamese hamlet called SW18
Ravaged by peace and the brutal schoolchildren
W. Pater came about the cooker
At some stage an exhaustive life of E. Pound
Appeared. A peasant named Confucius was raped
By the brutal intelligentsia. A large man came
About the rates, a book came about the rapes, a
Large book came about Shakespeare from Voltaire
To Ungaretti. An index slipped, trouble with
An appendix, my teeth hurt. At one point (I am
Almost sure) a book reviewer died at his desk
Whose name will be found at the bottom

In the Basilica of the Annunciation

A custodian, or guard, or butler
Reproves a youngish couple, their arms
Looped loosely, chastely round each other:
'This is no place for love.'

I ask myself
What this place is a place for.
This place is a building built by all the nations

This place must accommodate
(Without offence to any)
All the nations.

Its glories are the glories of the
Ideal Home Exhibition; the ark
Before the animals spilt drool and
Dung on its fragrant planks; a
Palace completed on the eve of revolution.
Many mansions do not make a house.

Is it a fact that in the streets
Camels are skipping through the eyes of needles?
When will the butler conduct us to the altar?

Coffee and biscuits will be served in the patio.
But first observe the Virgin on the walls
Figured in the physiognomy of every nation.
As for example
This elongated Japanese madonna in kimono
With an infant Jesus squatting on the air
At the level of her traditionally obliterated breasts.
They have the sly look of a city whore and her
Dwarfish ponce. But the nations must have their say.

This is no place for art
This might be the place for a cocktail bar
This is no place for little children.

Who shall roll away these stones?
Can any good thing come out of Nazareth?

Give me Manger Square
With its souvenirs, its Flowers from the Holy Land
That I found in my first bible.
Give me the gnarled and bashful Arab
Looking for something shabby enough to kiss.
Give me the disconsolate hippies
Competing for lifts with military heroes.
Or give me the Japanese lady
(For with God nothing shall be impossible)
Come down from the wall and telling me:
'This is no place for love, but I know
A handy cowshed.'

I blame this basilica for my sinful thoughts
This is no place for me.
The reproved couple, their empty arms
Dangling at their sides,
Have been married for quite some time –
If it makes any difference.

Along the River

They had pulled her out of the river. She was dead,
Lying against the rhododendrons sewn with spider's thread.
An oldish woman, in a shabby dress, a straggling stocking,
A worn, despairing face. How could the old do such a thing?

Now forty years have passed. Again I recall that poor
Thing laid along the River Leam, and I look once more.

They have pulled her out of the river. She is dead,
Lying against the rhododendrons sewn with spider's thread.
A youngish woman, in a sodden dress, a straggling stocking,
A sad, appealing face. How can the young do such a thing?

How Many Devils Can Dance on the Point . . .

1

Why, this is hell,
And we are in it.
It began with mysterious punishments
And the punishments led to the crimes
Which are currently being punished.
The more rational you are
(What you have paid for
You will expect to obtain
Without further payment)
The less your chances of remission.
Only the insane and saintly
Who kiss the rod so hard they break it
Escape to a palliated hell.
For the rest, why, this is it,
And we are in it.

111

2

Then what of those
Whose punishment was such, they
Never lived to carry out their crimes?
Children, say,
More than whose fingers were held
For more than a second in more than
The flame of a candle;
Though not exclusively children.
(No need to draw a picture for you:
The chamber, the instruments, the torture;
Forget the unimaginable, the
Imaginable suffices for present purposes.)
If the other was hell
Then what is this? –
There are gradations of Hades
Like the Civil Service,
Whereby the first is paradise
Compared with the last;
And heaven is where we are
When we think of where we might have been.
(Except that when we think,
We are in hell.)

3

Can this be heaven
Where a thoughtful landlord
Locates the windows of his many mansions
To afford you such a view?
(The chamber, the instruments, the torture.)
Can it be
The gratifying knowledge of having pleased
Someone who derives such pleasure
From being thus gratified?

4

Moves, then,
In a mysterious way . . .
Except that –
Lucid, strict and certain,
Shining, wet and hard,
No mystery at all.
Why, this is hell.

Terminal

A small boy, four years
Or so of age,
And tired and confused,
In a noisy, crowded building,
His ears still hurting
From some mysterious ailment.
He trails behind his parents,
Tired too, if less confused.

Then the people all take sides,
Like in a game,
His father joins the Caucasian file,
His mother the Other.
Which team is his team?
He hears them talking,
His English father, Chinese mother,
And the man who owns the building,

Who rubs his head:
'There's this queue and there's that queue,
There isn't any third queue.
I don't know what to say!'

Neither does the little boy,
He is tired and confused.
In front of him the two queues stretch away,
There isn't any third queue.

Metropolitan Water Bawd

The use of a hose
For non-domestic purposes
Can be quite costly.

And even more so
If an automatic sprinkler is employed,
Which goes without saying, and similarly
A perforated hose.

The hose should be held in the hand throughout
And the fittings must comply with the byelaws
And statutory requirements of the Bawd.

When a hose is in use, the flow
Should not be left to run unattended,
Otherwise you are liable.
The apparatus should be supported
Clear of the ground, using both hands
If necessary. Otherwise you are liable.

For remember: the conditions applicable
To domestic usage at private residences
Do not apply to non-domestic purposes
(Whether in private residences or public places)
Involving a hose, tube, pipe, sprinkler
Or whatever endearment you choose to employ.

For such usages there will be extra charges
Which may be paid in advance or subsequently
Or both.

But if you no longer desire to use a hose
Or if changed circumstances have rendered
Continued use inapplicable
Please delete the charge from your account
By detaching the portion and duly returning it.

An Ew Erra

The typeriter is crating
A revlootion in peotry
Pishing back the frontears
And apening up fresh feels
Unherd of by Done or Bleak

Mine is a Swetish Maid
Called FACIT
Others are OLIMPYA or ARUSTOCART
RAMINGTONG or LOLITEVVI

TAB e or not TAB e
i.e. the ?
Tygirl tygirl burning bride
Y, this is L
Nor-my-outfit
Anywan can od it
U 2 can b a
Tepot

C ! *** stares and /// strips
Cloaca nd † -
Farty-far keys to suckcess!
A banus of +% for all futre peots! !
LSD & $$$

The trypewiter is cretin
A revultion in peotry
" "All nem r =" "
O how they £ away
@ UNDERWORDS and ALLIWETTIS
Without a .

FACIT cry I! ! !

Another person one would like to be

Is a 19th-century composer of
Masses for the Dead.
God knows, one has the emotions anyway
One might as well believe in them.

No call to concoct a plot
No need to write the words
No lack of occasion –
There are masses of dead.

Once I wished I were an old Chinese gentleman
Glimpsed in a Chinese restaurant
Amid masses of Chinese relatives –
With the years one's ambitions grow humbler.

To a War Poet, on Teaching Him in a New Country

They talk about you as if you were alive –
To hear their remarks and profit thereby.
A compliment of a kind,

Though some of them think you frivolous
To stick a poppy behind your military ear,
And others complain of excessive sarcasm –
You are cynical about the war

(A suspicion persists, *re* those poppies,
That you were playing around with narcotics,
Perhaps to help you drop off again.
Court Poetic for you, my man,
Ten days Confined to Lyrics).

Worse, you are cynical about your mates,
Preferring rats to people, even 'haughty athletes'.
So much for Western humanism! And their Olympics.

While some find you equally unkind to the rat.
You accuse it of 'cosmopolitan sympathies'
And a paucity of proper rathood,
And you threaten to have it shot.

Practically all of them object to your attitude
Of boredom. 'The same old druid Time as ever,'
You mutter as you yawn across the parapet.
Poets are allowed to be boring,
But not to be bored –

Least of all when war is the question.
For these young people (government bursars,
Bonds to the whims of the Public Service Commission)
Have strong feelings on the subject of war,
An event which occurred between 1914 and 1918
In Europe.

They are very much against it,
You will be pleased to hear.

High-mindedness of an English Poet

I see, dear, in this book
That yet another bard
Will teach us how defeat
Is proper cause for pride
And somehow even sweet,

And that the virtuous man
Does not retaliate
But averts his bleeding eyes
Not to acknowledge hate.

Job is the case he cites
Whose readiness to sing
Under the frequent scourge
Was a fine and sacred thing –
But God was living then,

And you and I, my dear,
Seeing the bad go free
The good go by default
Know more than one sole state
Where this sweet bard would be
Appointed laureate.

Public Bar

Why are the faces here so lined?
Have they ever borne the pains of
Poetry? Or the strains of music.
Their hollow eyes have never searched a
Sombre canvas.

Their souls not scorched like ours
By burning issues. Or their cheeks
Trenched by the tears of things. No
Complex loves or losses wrung their hearts
Like ours.

Why do their faces look like this,
Carved through centuries, whole histories
Etched in their skin? Like works of art
Themselves. How did they steal
Our faces?

The Dispossessed

I will bare my bankruptcies
With a wealth of detail,
A manifest of absences
Will I then present to you.

I'll anatomize me next
Limb by limb, by inches,
And sing like chic couturiers
My multifarious bareness.

You will profit largely
From my catalogue of losses,
In my airless hovel
Shall arise a palace.

Things which *are* not –
They shall be there, every jot.
You will watch my separations
In unbroken copulation.

For these crazed ejaculations
Conjure up sheer sanity,
And these crumbled lifeless seeds
Conceive in you rich husbandry.

Till my frigid fingers shrivel
In the fire of my recital,
My tongue grows thick with love again,
I am possessed now, I fall dumb.

The New Year Too

The new year too is cruel. How long
Lord Mayor, the dustmen of my soul
Have been on strike! The snow too soon
Dissolves, and now the street's disgraces
Are released anew. What churlish
Beadles roll the carpet back?
The furthest suburb is not safe.

THE TERRIBLE SHEARS

Scenes from a Twenties Childhood

To my Mother

Little Buddhas

We had to keep our coal out at the back;
They wouldn't give us a bath.

I wondered why the Lotus Position seemed familiar –
It was how we crouched in the copper on bath night.

Uses of Literacy

When not in use, the copper
Was lined with old newspapers.
I was an early reader
(The one art that costs no money),
I could read upside-down –

Adverts for objects of
Unimaginable use, or the
News of the great world.
We didn't expect to understand
Such things.

The Nights

Unsure in summer,
You lie in bed in the light and feel you must be
Sick. Sheet lightning spoils the later night.

Safer in winter,
The oilstove throws its gently trembling patterns
On the wall. Accord of warmth and light and night.

Night-Light

In childhood there was this dream,
Recurrent and unyielding –
I was in a dark mill, or I was the mill,
The grindstones were grinding, with nothing
At all to grind, they groaned and groaned,
And when I woke, the dull heavy sound
Went on. I would wait for it to stop.
But in the end I had to call my mother.

Growing Up

I came to know this later dream
Like the two-times-table.
                    I was crossing
The Suspension Bridge. When I
Got half-way, I dropped to the ground,
Shrank paper-thin and slipped between
The rails. The water rose to meet me.
As I hit it, I woke.

'Soon be over,' I told myself, as I
Began to fall. Then as I neared the spot.
Then 'Soon be over,' I said, as soon as
The dream began. I was getting too old
To call for my mother.

Minorities

In the early Fifties we were the only
Foreign family in the hamlet of Okamoto.
At times an old crone would recoil
In the market, a baby howl with terror.
But the natives were very kind to us.

In my childhood there was only one
Black family in Leamington.
The natives were kind to them,
They were better-behaved than the Irish.

Respected Members of the Community

Mind you, this blackie was a postman;
As was my father, a mick.
Postmen were saintly figures in those days.
They were paid with cups of tea and kindly words.

Two Bad Things in Infant School

Learning bad grammar, then getting blamed for it:
Learning Our Father which art in Heaven.

Bowing our heads to a hurried nurse, and
Hearing the nits rattle down on the paper.

And Two Good Things

Listening to Miss Anthony, our lovely Miss,
Charming us dumb with *The Wind in the Willows*.

Dancing Sellinger's Round, and dancing and
Dancing it, and getting it perfect forever.

But Once a Year

The cracked oilcloth is hidden
By knife-creased linen.
On it bottles of Vimto squat,
A few flakes of browning tinsel
Settle. It is Christmas –
Someone will pay for this.

Jingle Bells

Our presents were hidden on top of the cupboard.
Climbing up, we found a musical box, in the shape
Of a roller, which you pushed along the floor.

This was for our new sister, she was only
A few months old, her name was Valerie.

Just before Christmas (this I know is a memory
For no one ever spoke of it) the baby quietly
Disgorged a lot of blood, and was taken away.

The musical box disappeared too,
As my sister and I noted with mixed feelings.
We were not too old to play with it.

A Glimpse

She was bending over the kitchen sink,
Milk, warm and unwanted, draining away,
Milk mingling with tears – or so I now think.

Where Did Dad Spend Christmas?

Christmas was always a bad time.
My father was on country delivery,
And either he broke his wrist when
Swinging the engine, or the van skidded
Into a ditch with him inside it.

(Once he only suffered from shock.
He ran over a large dog – 'it lifted the van
Clean off the ground' – but looking back,
He could see nothing on the road but snow.)

The hours passed and there was no sign
Of Dad. 'We'll let you know as soon as
We hear anything,' said the kindly Post Office.
We began to think he was Father Christmas.

Another Christmas

Another Christmas was coming.
Father thought of a way of enriching it for us.
He recalled that on the Somme
He had carried from the battlefield
A wounded officer by the name of Crawford.

(It was now a household name.)

He wrote to the gentleman in question,
Mentioning the not so distant incident
And the coming of Christmas. In return
There came a free packet of assorted biscuits.

They were consumed, and no doubt enjoyed,
Though felt to be less than they might have been.

Little Angels

Look through those family albums:
Each one contains a paradox.

Among recurrent characters
In careful suits and frocks
Or jacketless and jolly –
Sudden, unaccountable,
An alien apparition.

My mother, she could show a creature
Rosy in lace and velvet,
By-blow of some queer artist.
'A little angel!' Yes indeed.
I wouldn't have been safe on the street.

Where did all that finery go?
What became of those smug pretties?
They make their one appearance,
Then the Box Brownies take the lot.
Did the plague of Egypt wipe them out?

Somewhere there'll be a picture of him,
My idiot cousin,
Looking like the baby Plato
Or Infant Jesus.

Early Discovery

My sense of the superiority of women
Was confirmed at the age of seven.

My young sister was leaning against
The cast-iron railings of the balcony of
The third-floor room in which we lived.
The railings began to fall into the street,
The child began to follow.

My father and I were transfixed.
My mother, though hampered by a bread knife in the right hand,
Flew out and pulled the child back with the left.

Then my father had to go down and apologize
To a man in the street whom the cast iron had just missed.

Anglo-Irish

My father claimed to be descended from a king
Called Brian Boru, an ancient hero of Ireland.

My mother said that all Irishmen claimed descent
From kings but the truth was they were Catholics.

We would have preferred to believe our father.
Experience had taught us to trust in our mother.

Jephson Gardens

Two small children in the Gardens on Sunday,
Playing quietly at husband and wife.

How sweet, says an old lady, as she sits on
The bench: you must surely be brother and sister?

No, says the boy, we are husband and wife.
How sweet, says the old lady: but really you are
Brother and sister, aren't you now, really?

No, says the boy, trapped in his fantasy,
I am the husband, she is the wife.

The old lady moves off, she doesn't like liars,
She says. She doesn't think we are sweet any longer.

Flowers

The town was proud of the town's gardens,
People came from all over to view them.
I was taken there on Thursday afternoons
Or Sundays, when admission was free.

You couldn't play games in the Jephson Gardens,
Except for miniature golf at sixpence a round.

The aesthetic sense lay stillborn in me,
Those masses of flowers did nothing to me.

Later, having reached the Pathetic Fallacy,
I looked at them more closely: they were emblematic
Of something, I couldn't make out what.

On Sundays those masses of flowers pressed round me,
Muttering, muttering too softly for me to hear.
Their language, in my inept translation,
Was thick with portentous clichés.

I never learnt their true names.
If I looked at them now,
I would only see the sound of Sunday church bells.

Uncle Jack

Though all that time he never left the house
Uncle Jack got more and more weatherbeaten.
He was crippled with rheumatics
From building bridges over rivers.

He was the bad-tempered uncle,
His smoke-stained cottage the ogre's castle.

Next to bridges, milk was his favourite rage.
He wouldn't take a drop of it, he had worked
On a farm, he knew what went on there.
Uncle Jack on milk wasn't fit for kiddies.

They: Early Horror Film

Our pipes froze last winter
Because THEY took all the heat out of the air.

Not but what THEY are more to be avoided
Than envied.

Remember what THEY did to Joe Walters down the road –
Mucked him about properly.

THEY put Tom Binns in prison for stealing registered letters,
THEY drove up in their Sunbeams and caught him in the act.

THEY're a funny lot, but clever with it.
THEIR maker gave them fancies and the means to gratify them.

THEY drink blood after saying grace in Latin.

That Mrs Tooms opposite is stuck-up: she sits at
Her window drinking blood from a teacup, but I bet
She doesn't know any Latin.
Latin isn't for the likes of Mrs Tooms,
Nor is blood.
Even if her son is doing well in the Town Clerk's at Folkestone,
Or so she says.

Sometimes THEY try to get round you,
THEY come bearing scholarships.

Keep out of THEIR way, child!
Nothing but shame and sorrow follow.

The divel! Here THEY are, at the door –
Don't open it till I've put something decent on.

'Yes sir, no sir,
We wouldn't know anything about that, sir.'

THEY didn't get anything out of *me*!

Shades of the Prison House

How many remember that nightmare word
The Workhouse? It was like a black canal
Running through our lives.

'Old Mrs Povey has gone to the Workhouse.'
'You'll end in the Workhouse if you go on like that.'

It was shameful to end in the Workhouse.
Shameful to have a relative in the Workhouse,
The worst shame of all.
Such shame was always possible.

Even children came to dread the Workhouse.
Other times, other bogymen.

Patience

Patience was another potent word.
Patience was a virtue – in effect
It seemed to be the only one.

Certainly a virtue for virtue's sake,
There wasn't the slightest suggestion
It would lead to anything beyond
A respectable death, the undertaker paid.

Patience was our favourite game.
My mother is still addicted to it.

The word used to make us mad –
Patience, patience all day long!
Getting mad would lead to breaking plates
Or tearing one's clothes, that was all.

(And when we were short of cash
Who was the one to go short of food?
The patient one.)

Putting up with things
Was a speciality of the age.
You couldn't change things, and
Trying to would only make you miserable.

Nowadays clothes are manufactured torn,
There's hardly any china left.

It's not surprising that these days
No one puts up with anything, and
Patience is the rarest of words, and
Change the most vulgar.

Happiness

Yes, of course there were happy times,
It was not a succession of disasters.

(Once we went to Weston-super-Mare: the sea
Had apparently retreated to Leamington.)

The happiness you must take as read,
The writing of it is so difficult.

Playing

As kids we had great fun
With my granpa's Bath chair, till
He came shouting to chase us away.

In Siam I watched urchins
Playing about a pedicab, till
The driver chased them away with curses.

A lowering way of earning one's rice;
For children a treasure trove.

Facts of Life

I had two white mice –
Then I had scores of baby mice,
Naked and pink little things, dead or
Alive or half-eaten.

Those two small mice did me more harm
Than all the pornography in the world.
My father took them away.

The dog went too.
Him we liked. But he was a creature
Of the open spaces, he needed
Long hours of exercise in the park.

He went back to the country. I hope
That was where he went.

War Game

When the soldiers lost their limbs
They were turned to wounded soldiers

Put to nursing, a farmyard milkmaid
Carried in her pails the limbs of soldiers

A section of the rails was missing: here
The track blew up, hurting the soldiers

A general might have to ride a hippo
Commandeered from Noah's Ark

A wooden tiger carry off the wounded
The milkmaid and her pails as well

The more things went to pot
The more authentic the whole thing got.

Rude Health

Brimstone and treacle (not bad that),
Powders disguised in strawberry jam
(Even now I can't touch strawberry jam),
California Syrup of Figs and Parrish's Food,
Senna tea (it spoilt me for the Chinese kind),
Cascara sagrada (which I now perceive

Bears a most impressive etymology),
And something peculiarly horrible
Which we should have been given more often
Except that no one could pronounce its name –
Ipecacuanha Wine.

Such the incessant medication of our childhood.

Since when I have needed nothing beyond an
Occasional aspirin.

Young Criminals

'The same to you with knobs on,' said the first.
'The same to you with spikes on,' said the second.
'The same to you with balls on,' said the third.

The master heard us. He was a just man.
The first of us got one stroke on each hand.
I got two strokes on each hand.
The third got a fearful beating –
He was a dunce, he smelt, his name was Bugg.

(About the time I left Clapham Terrace Elementary
He was put away for showing his thing in the street.)

Training

How docile we were, how orderly! Empire Day,
Armistice Day, and all that religious instruction!
They were training us to die for something –
It meant nothing, only holidays and queer emotions.

Forty years later, walking in Canton, I encounter
A mass of orderly children – they are listening
Intently, with every sign of agreement,
To a horror story about red-haired imperialists.

I slope past fearfully. But to them I'm no more
Than a comical flower in this well-kept park.
Keeping one's eyes on teacher is far more important.
As yet they haven't learnt to connect.

A Sign

At an early age I achieved a sort of
Puzzled fame in the family circle.

I retrieved an old broken-backed
Bible from the dustbin and bore it
Back into the house. No doubt
With a scandalized look on my face
Though not wishful to chide these
Blasphemers against God's Word.

At that tender age I couldn't bear
To see printed matter ill treated.
I would have subscribed to the ancient
Oriental taboo against stepping light-
Mindedly over paper inscribed with characters.

It was read as a sign. The child
Is destined to become Vicar of the Parish Church!
He has rescued Religion from the scrap-heap.

Now I could watch unmoved the casting
Of hundreds of books into dustbins.
But two of them I think I should still
Dive in after – Shakespeare and the Bible.

Sunday

My mother's strongest religious feeling
Was that Catholics were a sinister lot;
She would hardly trust even a lapsed one.
My father was a lapsed Catholic.

Yet we were sent to Sunday school.
Perhaps in the spirit that others
Were sent to public schools. It
Might come in useful later on.

In Sunday school a sickly adult
Taught the teachings of a sickly lamb
To a gathering of sickly children.

It was a far cry from that brisk person
Who created the heaven and the earth in
Six days and then took Sunday off.

The churches were run by a picked crew
Of corny actors radiating insincerity.
Not that one thought of them in that way,
One merely disliked the sound of their voices.
I cannot recall one elevated moment in church,
Though as a choirboy I pulled in a useful
Sixpence per month.

Strange, that a sense of religion should
Somehow survive all this grim buffoonery!
Perhaps that brisk old person does exist,
And we are living through his Sunday.

Uncertainties

Our folk didn't have much
In the way of lore.

But I remember a story,
A warning against envy
And also against good fortune,
Too much for our small heads –

About a lucky man called Jim
(My uncle in Dublin I used to think,
But he was Sunny Jim)
And his friend who envied him.
Jim had the luck
He married the girl they both of them loved
And his friend envied him.
Then Jim died, and the friend
Married his widow. And then
The friend envied lucky Jim,
Asleep in peace in the churchyard.

When Granpa wasn't pushing old ladies
Through the streets of the Spa
He would cut the grass on selected graves.

Sometimes we went with him. Dogs
Had done their business on the hummocks.
The water smelt bad in the rusty vases.
The terrible shears went clack clack.

It was too much for our small heads.
Who was it that we mustn't envy –
The living or the dead?

The Pictures

Threepence on Saturday afternoons,
A bench along the side of the hall –
We looked like Egyptian paintings,
But less composed.

Sometimes a film that frightened us
And returned at nights.
Once *Noah's Ark*, an early talkie
We took for non-fiction.

Cheapest was the home kino.
Lying in bed, you pressed on your eyes,
Strange happenings ensued.

But the story was hard to follow
And your eyeballs might fall in.
Fatigued, you fell asleep.

Spoilt

How well we were catered for!
No wonder we lost our teeth.
Chewy locust, thick strong liquorice sticks,
Aniseed balls, bull's-eyes, and sherbet . . .

Later in my prime, at elegant parties in the
Orient (but where was the sherbet? Where
The locust?), I met caviare and smoked salmon
And various oriental delicacies.

They were no sort of substitute.
But happily I found a new strong taste,
Easy on the teeth too,
Whisky, gin, brandy and ginger.

Pleasures of Reading

Aged ten or so, I read *The Well of Loneliness*
(How did it enter the house?).
The book left an impression on me
Both indelible and indecipherable.

Aged forty, I reread *The Well of Loneliness*
And could not recognize it.
Somewhere there must be another book,
The Well of Loneliness I read at ten or so.

Escapism

'That was a miserable poem you wrote about
The Black Country,' said an old WEA faithful
In Cradley Heath. 'It's cheering up we need.'
At the next meeting I tried to put him right
On this central literary issue.

Not helped by recalling that in the Twenties
And later our staple diet was *Red Letter*
And Ethel M. Dell and *Old Moore's Almanac*,

And that if you can escape for a moment
And a moment's escape is all you can manage,
No one has the right to forbid you.

1,000 Useful Things
To Do About the House

How creative we were in those days!
We made things out of nothing
With our own hands.

In a week of evenings, for instance,
Several miles of cork-wool . . . Enough
Papier-mâché bowls to stock the V & A . . .
Trays made from melted gramophone records
Sufficient to equip Joe Lyons . . .

Now people sit and watch television.
Which is often quite instructive –
You can even listen to talks about the
Conspicuous creativity of the old days.

Preparing for Life

Inclined to pedagogy, my favourite pursuit
(Indeed I conceived it a duty) lay in the
Education of my young sister by means of
A small blackboard and an overbearing manner.

It was not that my sister was unteachable,
Rather my methodology was out of date.
The lesson always ended with the class in tears
And the teacher summoned before God the Father.

Whatever Sex Was

It was the two sisters next door
Quarrelling over their husband, or
Their drunken husband punching them.

It was the trouble that some woman
Was in, a mysterious trouble that
Could only be talked of in whispers.

137

It was the man who frightened my
Little sister, and whom my father
Searched the streets for, for hours.

Or the man who got angry with me
In a public lavatory, and followed me
Into the street with inexplicable curses.

It was men fighting outside the Palais.
It was crying, or it was silence.
Whatever sex was, it was another enemy.

A Grand Night

When the film *Tell England* came
To Leamington, my father said,
'That's about Gallipoli – I was there.
I'll call and see the manager . . .'

Before the first showing, the manager
Announced that 'a local resident . . .' etc.
And there was my father on the stage
With a message to the troops from Sir Somebody
Exhorting, condoling or congratulating.
But he was shy, so the manager
Read it out, while he fidgeted.
Then the lights went off, and I thought
I'd lost my father.
The Expedition's casualty rate was 50%.

But it was a grand night,
With free tickets for the two of us.

Bad Day

Talking of sticking her head in the gas oven;
Humming 'There are many, sad and dreary';
Putting out bread and a basin of dripping.

A Difference

To drown yourself in the cut
You would have to loathe yourself.
A person with any self-respect
Made use of the river. The town
Was named after the river.

Euphemisms

After the main entry in clinical Latin and Greek
Which I got by heart in order to dazzle my schoolmates,
The Certificate abruptly changed its tone and remarked
That a Contributory Cause of Death was Septic Teeth.

The oddest thing, however, was to find that the Deceased
Was known as George Roderick. Perhaps this was clinical
Language too. No one ever called him anything but Mick.

Insurance

One spot of cheer in the Midlands gloom
Was our Dublin uncle
Who sent us shamrock each St Patrick's Day
And ebullient letters
On paper headed THE PHOENIX ASSURANCE COMPANY LTD.
He was the family success
(His photograph looks like Mickey Rooney
He could sign his name in careful Gaelic)
He moved in the corridors of power.

He came across for his brother's funeral
Pensively noting the widow and orphans at the graveside.
Something had to be said about our insurance.

He had borrowed, he intimated, the company's notepaper.
Faith, he worked there, he was a janitor there
He moved in the corridors.

Romantic Ireland was dead and gone.

The Lodger

Our council house would be gone as well
Unless we found a way to pay the rent.

We took a lodger, a large schoolteacher
Saving up to buy a house and marry,
His talk was all of saving.

And then his fiancée, a large girl,
Took to staying with him at weekends,
In what was once the parental bed.

It wasn't that we thought it sinful
Exactly (they were going steady),
But a slight malaise infected the house.
Perhaps it was fear of a bad name,
Or a sense of being diddled.
He was only paying rent for one.

No more lodgers, said my mother,
As the couple left for their honeymoon.

Iron Horse

It must be admitted that Granpa
Never had a day's illness in his life
Till the time he went to the station
To see a niece off.

As the train swept in, he fell to the ground
Dead. The relicts considered the idea
Of suing the Great Western Railway for
Creating an excessive wind with fatal results,
But abandoned it.

The Bath chair went back to the owners.

The Exception

Granpa was a sturdy exception.
Sickness too was different in those days,
People tended to die of it.

My father might be said to have invented
Lung cancer (and without benefit of smoking).

They were carried off one after another.
Such contraction of the family circle!

We noted, my sister and I, that flesh
Was weaker than the couch-grass in the yard.
It hardly seemed worth continuing with school.

Geriatrics

I got on well with Granma.
Ours was the prescriptive relationship:
We used to play crib for hours, she gave me sweets,
She defended me, her I failed to defend.

It didn't worry me that she was getting
Troublesome. If she wandered in her mind a little,
So did I. Her husband was dead,
So was my father. The house had to be vacated.

We couldn't look after her, we were going
To look after a troublesome old man
Who at least had a house we could live in.
My eldest aunt declined to take her.

She would have to go to the Workhouse.
The worst thing was, they told her
She was going to a nursing home for a while.
They even ordered a car.

She had to be pushed into it.
As the car was moving off, I heard her
Shout with a dreadful new voice:
'I know where you're sending me,
You're sending me to the Workhouse!'

141

She was found to be deranged on arrival,
And they sent her on to another place.
So she didn't go to the Workhouse after all.
She died soon after.

Early Therapy

Granma doddered a bit,
But she was my friend.
Perhaps it had to be done,
Did it have to be done like that?

It started me writing poems,
Unpleasant and enigmatic,
Which quite rightly no one liked,
But were thought to be 'modern'.

It Is Poetry

As Leverkühn began his last address
To the cultivated ladies and gentlemen
There assembled,
They were highly bewildered.

Till one of them cried,
'Why, it is poetry! One is hearing poetry!'
Thus relieving them all immensely.

But not for long –
As the composer's friend noted –
Alas, not for long did one think so!
They were hearing about damnation.

It sent the speaker mad.
The listeners it sent home indignant.
They had expected an artistic soirée.

Always Learning

The gym teacher was big, handsome and
Dashing. He pronounced *tooth* as *tutb*
And *food* as *fud*, which much intrigued us.
He talked of rugger like a lover,
And rode a motorbike, big, handsome
And dashing, which we much admired.

He broke a leg in an accident
On his bike, and couldn't perform
The exercises he required of us.
He described them in words, like an elegy,
A lovely man! We groaned twice over,
For ourselves, and our disabled hero.

The months went by, he ceased to limp, but
In the gym was still the thwarted cripple.
We groaned for our sweating selves alone.
Do as I say, not as I do:
We had gained in worldly wisdom,
We had lost an admiration.

Scholarship Boys

How docile the lower orders were
In those distant days! Having done
Unexpectedly well in the School Cert,
I was advised by the head to leave school
At once and find a job before they spotted
A mistake in the exam results.

And I almost did.

Weren't There Any Windows?

So my mother went to keep house for an old man,
A tyrant and a hypochondriac.
If he couldn't get what he wanted by shouting,
He got it by weeping.

143

He claimed he couldn't walk six yards to the lavatory.
'He killed his wife,' the neighbours warned,
'He will kill your mother.'

I resolved to get in first.
(Perhaps if I refused to fetch him his bucket
His bladder would burst?)

Then he became truly ill.
And one night the growth in his bowels burst,
And he died.

The stench was unexampled. My sister and I
Sat up till dawn with handkerchiefs round our mouths.
Later the fumigators came and sealed off the room,
And we packed our bags again.

A Step Up

The next one my mother kept house for
Was a good deal more civilized.
He was headmaster of the school I went to.

He said he would be a father to me,
He said it several times with emotion.
His subject was English.

He taught me to drive his car from a nearby garage
And leave it outside the house in the mornings.

My mother fell ill and had to go to hospital.
My sister took his boiled eggs to the headmaster
In the mornings, and often dropped them
At his feet. She wasn't at home with headmasters.

My mother went on being ill,
And he decided to marry the daughter of a
Prominent and well-to-do local citizen.

Alone in a classroom he told me
Of *Romeo and Juliet*,
Then fell into a dream, while I sat there
Embarrassed.

He lost a family but gained a wife.
Some years later he died without issue.

Said the Straw

'It's the last camel,' said the straw,
'The last camel that breaks our backs.'

'Consider the lilies,' he said,
'How they toil in the fields.
Six days they labour,
On Sunday they wear their best clothes.'

'Ask why the violet sickened,
The pale primrose died unmarried,
And the daisy lies in chains,'
He said, 'All grass is flesh.'

'But let us hear no more,'
So said the straw,
'About the sorrows of the camel
With its huge and heavy feet.'

A Much Later Conversation

'Your father –
Did he die in error?'

Well, I suppose you could say that.
He was laid up for several months,
Then they said he was well, and
He went back to work. The next year
He was ill again, and this time he died –

'Yes, but what I mean is – '

Well, a widow's pension wasn't much.
We put it to them that getting gassed
In France had buggered up his lungs.
But the War Office wouldn't wear it,
They observed that the war had been over
For some time . . .

'I'm sorry, but I didn't mean – '

Oh yes, we had a priest in,
The first we'd ever seen at close quarters.
He gabbled in Latin,
And no one could understand him.
But at least –

'I'm sorry, but you've misunderstood me.
I was only asking
Did your father die in Eire?'

No, he died in England.
But you're right, it may have been a mistake.

Knives

Sheath-knives fascinated me
(Less than guns, but they were out of reach).
I owned several over the years, horn-handled,
Sheaths so stiff you could hardly pull the knife out,
And blades you could never sharpen.

On one occasion, having failed
To pierce the skin of my breast,
I turned to sharpening my pencils.
The knife wouldn't even harm a pencil –
I had to laugh.

A Bookish Boy

On a later and more earnest occasion
I had recourse to literature.

In this novel a youngster
Rolled in the dew on the grass and
Went to bed and caught consumption
And died.

I was taken with the dignity of it.

There being no dew available
I soaked my pyjamas under the kitchen tap
And went to bed. After a foul night
I arose from a bed of steam
Exhausted but essentially undamaged.

Plainly my mother suspected an accident,
She was too considerate to press the point.

The Doom Game

In those days warnings were delivered
In good round terms. One took them
Seriously.
(Except for the criminal elements
And then a few of us who made
The sceptic Sixth Form.)

These days warnings have grown common.
Hardly a wall without one, hardly a passerby
But bears a warning in his hand or mouth.
Warnings comprise our chief entertainment,
The direr the dearer,
On the radio, in the press,
In verse, in prose,
In literary criticism.

Yet one thing remains constant –
The high spirits of the warners, their
Confident step. Fears for the future
Have never cramped their present pleasures.

And now as then,
The ones who suffer already, or
Those who lie in the path of disaster –
Their mouths twitch, they cannot
Get the words out. Mumbling at most,
'It's the same old story,' or
'You can't change human nature.'

Such joyless clichés butter no asparagus.

Learning to Hick and to Hack

'Much have I travelled in the realms of gold
for which I thank the Paddington and Westminster
Public Libraries . . . ' – Peter Porter

Behind official bricks
We found the township's casual treasure.

The Public Library
Handsomely stocked and not used to excess
By the public – it was almost like
Having a library of one's own.
Without it, some of us wouldn't have lasted
Much past adolescence.
Into what strange routes it led us,
What pungent semi-understandings!

Bravo, England of the Thirties!
Your smallest dullest town enclosed alternatives
To littleness and dullness.

Only once were we betrayed –
In the Reference Section, where we signed our names
For the Loeb crib of the *Fasti* or the *Civil Wars*.
Loeb was more complete than our school editions,
And our faultless versions gave the game away.
So did our signatures
When the teacher called on the librarian.
I went on rather liking Ovid, though.

The Poets

Wordsworth, Keats and Shelley –
How to picture what they meant then?
(Or the meaning that we lent them)
The critics do not tell me –
And I'm not eager to remember.
'The words on the page' came later,
When one could afford them
When one was stronger.

148

Times Change

One never complained of being misunderstood!
That in fact was what, if ruefully, one hoped for.
To be understood would have been calamitous.
Silence, exile and cunning were our devices.

So now the public young and their blatant banners
Astonish me. Such honesty, such plain speaking!
I find I am blushing with retrospective shame.

Second Thought

It was a short-lived shame.
At all events, the blush soon faded.

A sort of honesty is safe now. What's safe
Becomes an easy fashion. Then begins to die.

I think of some I knew in Asia,
And their tiny squeaks of protest. A
Big cat watched them, and they knew it.

I pay a privy tribute to those brave mice.
What would they think, I wonder, of these
Jolly marchers flanked by well-bred horses?

Religious Phase

'Take an unsteady schoolboy, his whole sense
is an unskilful posture of defence' – Peter Levi

It was the undefended one felt for.
On the third day
He arose from the dead and no doubt was
Well received at Heaven's gate.
He was on secondment. At no time
Was he ignorant of his state.

His ignorant bewildered mother
Was another matter.
In our street the pangs of labour
Were nearer than those of crucifixion.
Carpenters were useful, but
Every family required a mother.
The dirty end of the stick was known to us;
Nine months for a start
In an unskilful posture.

Because of the fear of Rome, we
Hadn't heard too much about her.
Our Church was run by married men;
They were minded to put her away privily.

As for those male toffs,
His ineffective entourage –
Johnny-come-latelies,
Made out of stained glass.

Difference of sex was no bar;
To an unsteady schoolboy
It could even have been a lure.

The Soul of a Schoolboy

A woman thrust her way into the house,
Desirous to save the soul of a schoolboy.

An obliging schoolboy, would do anything
For peace, excepting kneel in public.

But no, she would not go, she would not go,
Till crack on their knees they fell together.
His soul was lost forever.

Ugly Head

There seems to be a large gap
Somewhere about here.
If repression is at work
Then repression works efficiently,
In this sphere.

I don't remember learning about sex
In the school lavatories;
Though I remember the lavatories.

With a great effort I call up
Certain goings-on in the rear rows
Of the Physics class. I can't believe it.
That Welsh master was so sharp
You couldn't blow your nose
Without him glaring.

At one time or another
Some slightly special one or other –
But to kiss a girl
Would have seemed like criminal assault.
There was one called Pearl
Who would quote bits from Rosalind
In *As You Like It*, leaving me confused.
Once at a party I stepped heavily
On her hand, and was appalled.
In a strange way she seemed to like it.
I was glad to go home and study *The Prelude*.

It was homework and rugger; then
It was essays and walks to Grantchester.
Perhaps we were great Platonic lovers then.
Perhaps there is nothing to remember.

Class

I can't help it, I still get mad
When people say that 'class' doesn't mean
A thing, and to mention one's working-class
Origins is 'inverted snobbery'.

The wife of a teacher at school (she was
Mother of one of my classmates) was
Genuinely enraged when I won a scholarship.
She stopped me in the street, to tell me
(With a loudness I supposed was upper-class)
That Cambridge was not for the likes of me, nor was
Long hair, nor the verse I wrote for the school mag.

Her sentiments were precisely those of the
Working class. Unanimity on basic questions
Accounts for why we never had the revolution.

The Cure

Plainly there was a lot to be done.
The trouble was, mere survival
Took up so much time and strength.

When I went to Cambridge
I found the ills had been diagnosed,
And correctly too.

But I wasn't too sure of the cure,
And couldn't quite believe
Things had once been so much better.

Sentimental Journey

On a Sunday in 1968, on leave from Asia,
I set out to visit the house where I was born.

The town stirred with dust and papers,
Last night's vomit was scarcely dry.

It was cleaner than this in our day;
Perhaps we had less to throw away.

I found the street, a *cul-de-sac*
(Pudding-bag was what we called it).

Pakistanis stood there, one beside another,
Washing their second-hand cars.

They eyed the stranger not too kindly.
What could his business be in that *cul-de-sac*
But collecting payments on their cars?

Ancient Fears

Accustomed to dropping into
Kempinski's, Baudrot and Raffles,
Ashoka, Ichi-Riki, Hoi Thien Lao,
And the harbour bars of Alexandria –

I enter the Cadena on Victoria Parade:
The fine matrons still frighten me.

Large Mercies

I remember the schoolgirl under the bus,
Her bicycle lying in her blood,
And the driver in tears, saying over
And over, 'I'll never drive again.'

I remember too, her leg was amputated,
And when she passed her exams
The local paper announced it proudly,
And again when she married.

That means it wasn't a bad life.
No one was dragged out of bed by
Armed men. Children weren't speared
Or their brains dashed out. I don't
Remember seeing a man starve to death.

That's something we shouldn't forget –
That we don't remember things like that.

Acknowledgements

'But he has been equipped for hurdle-jumping;
so he merely dreams of getting-on, but somehow
not in the world's way.' – Richard Hoggart

The completion of this book is in the first place
A tribute to my typewriter and its associate, the
Parker Pen. I must also acknowledge with thanks
The kindly guidance of memory, and the ever-patient
Encouragement of the author by his close friend,
The author, to whom he is also grateful as at once
The inspiration of his studies and the seminal impulse
For his interest in himself. My labours have been eased
By the gracious hospitality of my desk and the unfailing
Support of my chair. I am indebted to the Wandsworth
Borough Council for placing valuable space at my disposal,
To the London Electricity Board for throwing light on
The long winter evenings, and to the learned society of
The *COD*. Without the research materials made available
By Life and Time, I should often have been at a loss.
Sincere apologies are offered to long-suffering friends
And relatives who would rather have lived in the present.
And finally I would record my appreciation of the
Grants-in-aid provided by Messrs Gordon, Haig and Watney,
And in particular my debt to the following Foundations:
Father, Mother, Midwife.

It takes a long time to learn a new language;
But one almost gets there in the end.

SAD IRES

Since Then

So many new crimes since then! –

    simple simony
    manducation of corpses
    infringement of copyright
    offences against the sumptuary laws
    postlapsarian undress
    violation of the Hay-Pauncefote Treaty
    extinction of the dodo
    champerty and malversation
    travelling by public transport without a ticket
    hypergamy and other unnatural practices
    courting in bed
    free verse
    wilful longevity
    dumb insolence
    bootlegging and hijacking
    jackbooting and highlegging
    arsenic and old lace
    robbing a hen-roost
    leaving unattended bombs in unauthorized places
    high dudgeon
    the cod war
    massacre of innocents
    bed-wetting
    escapism
    transporting bibles without a licence

But so many new punishments, too! –

    blinding with science
    death by haranguing
    licking of envelopes
    palpation of the obvious
    invasion of the privacies

fistula in ano
hard labour down the minds
solitary conjunction
mortification of the self-esteem
the Plastic Maiden
hat-rack and trouser-press
the death of a thousand budgerigars
spontaneous combustion
self-employment
retooling of the economy
jacks in orifice
boredom of the genitals
trampling by white elephants
deprivation of forgetfulness
loss of pen-finger
severance pay
strap-hanging
early closing
sequestration of the funny-bone
mortgage and deadlock

'Fair do's,' murmured the old Adam, 'I am well pleased.'
He had come a long way since he named the animals.

The Stations of King's Cross

He is seized and bound by the turnstile.

The moving stair writes once, and having writ,
Moves on.

At Hammersmith the nails
At Green Park the tree.

A despatch case which is well named
A square basket made of rattan
Which is a scourge.
The heel of the Serpent bruises Man's instep.

At Earl's Court a Chopper
New, flashing, spotless
It carries hooks and claws and edges
Which wound.

It is hot. Vee
Wipes her face. Cheek to jowl
She wipes the man's on either side.

Rather bear those pills I have
Than fly to others that I know not of.

He speaks to the maidenforms of Jerusalem
Blessed are the paps which never gave suck.

The agony in Covent Garden
He finds them sleeping, for their eyes are heavy.

The first fall, the second fall
The third fall.
And more to come.

A sleeve goes, a leg is torn
A hem is ripped.
This is the parting of garments.

They mock him, offering him vodka.
The effect is shattering.

He is taken down from the strap.
And deposited.

Wilt thou leave him in the loathsome grave?

Origin of the Haiku

The darkness is always visible
Enough for us to write.
We pass the time composing *haiku*.
It concentrates the mind –

Counting every syllable,
Revising, counting again.
A lot of thought goes into 17 syllables,
A lot of time, and (you might say)
A lot of pain.

157

Once a desperate faction
Proposed to bring in rhyme,
A trick for using up more time –
They lost by a large majority.
We are a conventional lot,
This is a conventional spot,
And we take some satisfaction
In writing verse called *free*.

In between we make up epigrams.
'Not to know me argues yourselves unknown',
Or 'What is else not to be overcome?'
The mind is sometimes its own place.

Such petty projects –
Yes, but even an epic,
Even *Paradise Lost*,
Would look puny
In hell, throughout eternity.
It's the taking of pains that counts.

Eternity

I can see it now,
Sitting there, next to the Doctor
Listening to Berlioz' *Damnation*
Then discussing it at length.

Latini will have something to say
You can be sure
On being ahead of one's time.

Don Juan on Mortuary Art
And the Burden of the Past.
The Old Man's recension of Milton:
Ancient Crux and New Light.

The Case of the Fallen Angel –
Was He Pushed?
A symposium led by Belial.

And the occasional guest speaker:
Jesus on The Harrowing –
Did It Really Work?

There is no end to torments,
The old can be used again and again.
It's in the other place
The programme is a problem.
Heaven help them.

The Wine List

' – And Paradiso? Is there a paradise?
– I think so, madam,
   but nobody wants sweet wines any more.'
                    – Eugenio Montale, trans. G. Singh

Not so fast, waiter.
If there are those who like sweet wine
And have earned the price of it,
Then they should have it.

Plonk will do for me.
If there's cork in it
Or lipstick on the rim,
I shan't make a fuss.
Some of us will be lucky to get vinegar
Pushed at us.

But for others
You had better be ready to serve sweet wine
In clean glasses, unchipped,
And without a speech.
Some customers are by definition
Right,
And do not require to be told
About a fine dry wine
Deriving from individually crushed grapes
Grown on a certain slope on a small hill
Overlooking a distinguished river.

Some of your customers
Have already been individually crushed.

159

They know dryness in the mouth,
A harsh taste at the back of the throat.
If sweet wine is what they fancy
You will give them sweet wine.

And there should be room on your tray
For ginger beer, orangeade, cocoa, tea
And even the vulgar vintage of colas.

In Cemeteries

This world a vale of soul-making –
To what intent the finished wares?

Is the ore enforced and fired through
Harsh mills, only to fall aside?

Who is this soulmaster? What say
Do souls have in their made futures?

We mourn the untried young, unmade
In small coffins. What of grown graves?

At times in cemeteries, you hear
Their voices, sad and even-toned,

Almost see the made souls, in their
Curious glory. If you are old.

Three Girls

Faith had a fatherless child,
Something had been misplaced.

Charity saw good in everyone,
Until her eyes gave out.

Hope became a chorus girl,
She springs eternally.

Who is the greatest of these?
They are equal. They abide.

R-and-R Centre:
An Incident from the Vietnam War

We built a palace for them, made of bedrooms.
We even tracked down playmates for them
(No easy job since prostitutes went out
When self-rule came). We dug a pool,
Constructed shops, and a hut for movies
With benches outside for the girls to wait on.
Serene House was what we called it.
We did our bit in that war.

Air America brought them from the battlefield.
We lifted the girls from the suburbs by buses:
Chinese, Indian, Malay, Eurasian,
Healthy and well-fed and full of play.

There were cameras in plenty, tape-recorders
And binoculars for the soldiers to buy
For the girls; for the girls to sell back
To the shops; for the shops to sell to the soldiers.

Serene House was near the varsity. The GIs
Strayed across the campus with Nikons and blank faces:
It was feared they might assault the female students.
They seemed scared of their own cameras.
They looked at nobody; nobody looked at them.
                         That violence down the road –
It was good for business, and we did our bit.
Otherwise it was a vulgar subject.

Once I found a GI in the corridor,
Young and dazed, gazing at the notice-boards.
The Misses Menon, Lee, Fernandez, Poh and Noor
Should report for a tutorial at 3 p.m.
Bringing their copies of The Revenger's Tragedy . . .
If Mr Sharma fails to pass his essay up this week
He will find himself in serious trouble . . .
The Literary Society seeks help in cutting sandwiches . . .

He was still there thirty minutes later,
A stunned calf. I asked if I could help.
He shrank away: 'Is it not allowed to stand here?'

The corridor was dingy, walls streaked with bat shit,
Somewhere a typewriter clacked like small arms.
'Is there . . . would there be a . . . library?'
One of the best in fact in South-East Asia. –
I offered to show him. He trembled
With a furtive pleasure. His only licence
Was to kill, to copulate and purchase cameras.

What sort of books would he like to see?
Outside in the quad he was jumpy,
As if unused to the open. He glanced behind,
Then whispered. Yes, there was something . . .
Did I think . . .

What could he be after? The Natural History
Of the Poontang, with Plates, by some defrocked
Medico called Aristotle? How to Get to Sweden
By Kon-Tiki through the Indian Ocean?

'Would they have anything . . .' A quick look
Round – '. . . by Cardinal Newman, do you think?'
I left him in the stacks, the *Apologia* in his hands,
He didn't notice when I went away.

Inside Serene House, in the meantime,
Girls galore (such lengths we went to!)
Lolled on the benches, played with binoculars,
Clicked their empty cameras, and groused.
The soldiers were happy to quit Vietnam;
Five days with us, and they were glad to go back,
Rest and recreation, they said, was too much for them.
We weren't surprised when the Americans didn't win.

Hands Off, Foreign Devil

Recounting softly some small
Tortuous oriental sorrow,
With dainty Chinese finger-tip
She crushes one by one the ants

Advancing on the office sugar.
Absorbed in her dilemma, trifling
And enormous, the typist deftly
Rubs the ants out one by one.

I turn towards the window,
Listening to her, but not looking,
Not lecturing. The country's hers,
Not mine. The ants too, I suppose.

Meeting a Person

              This here is a Person It is not for
Screwing or soft-soaping or scratching your back against
It is not for weeping on (it hates all kinds of wetness)
Nor will you be permitted to do things to it in the dark
Of a dance hall or a side-street or when it is not looking
Because it is always looking This is a young Person here
It will do what it is asked to do if it approves the request
But otherwise otherwise It has a mind of its own a body of
Its own and above all this Person has a person of its own
And of nobody else It holds views which are its own views
And not necessarily those of the majority or the minority
And all its views are pronounced and it pronounces them all
For this Person is fearless and not afraid to say so because
It believes in absolute honesty to its person and will readily
Exchange opinions on a variety of topics ranging from the
Pollination of orchids through the Poems of Keats to the
Five Sacred Appurtenances of the Sikhs It loves its family
But it is a Person and reserves the right to disagree quite
Radically with its father and mother and brother and sister
It is (it tells you again) a Person and a Person is something
Which many people are not whether in the dark of a dance hall
Or the bosom of a family it remains a Person and is only
Concerned to do what this Person considers right and not to do
What this Person considers wrong

              And naturally one approves of people
Being Persons especially in the Eastern society to which
This person hardly seems to belong and one is obliged to
Say so and what's more in a tone of admiration For it is
True that not every person appears to be exactly a Person

And it is splendid that this person should stand on its
Own two feet (does it have only two feet? One doesn't dare
To seem to stare) for *being a Person* in fact is the burden of
Much of one's day to day teaching from year to year so truly
One ought to be happy and pleased to hear that this here is a
Person at last Yet Goodbye one says smiling feeling a little
Diminished and unsure of one's own little person and not too
Proud as one thinks of one's one and only little joke when
One asked Did it say it was a Parson? And one fears that
In the dark of a dance hall or the bosom of a family or among
The limbs of strangers this Person may be going to be a rather
Lonely person Yet Goodbye one says and perhaps after all
God will be with this Person

Home and Colonial

Henri Rousseau's 'Tropical Storm with Tiger'

I'm not one of those simpletons who believe
That if only they had a larger TV screen
They would be able to see the naughty bits.
But if that picture were a few inches longer,
Here on the right-hand side, I mean – then
In fact you would see – not a naughty bit –
You would see me.

Sexual behaviour does exist in the tropics –
Oh indeed – but it's relatively invisible.
It doesn't go on in public. And it wouldn't
Even if there weren't a storm, even if
The jungle weren't so full of spiky things.

Public sex is less sex than public, I reckon.
Like that young couple in the Underground
The other night. They weren't doing anything,
They were simulating it. In my day
We used to dissimulate. And likewise I doubt
This notion that a wider screen creates
A broader mind. What you can see is never
The interesting part. Though of course
I'm not referring to a gentleman like you
Looking at a picture like this
In a reputable gallery.

Imagination is allowed some latitude,
I know (though, as it happens, this painting
Doesn't get enough), but all the same . . .
The jungle's not half as pretty as it looks here,
Untidy at the best, storm or no storm.
The bougainvillaea was tatty and blotched,
Not right out of a hothouse. It was gloomy –
That's another thing about jungles – and
The lightning had that lost air it always has
In those parts. Fumbling around for something
To get a grip on, like a roof, a chimney
Or a golf-club.

But the tiger – Frenchy's hit it off to a T!
Scared stiff, what with its tail behind, which it
Took for a flying snake, and in front – a hairy
Red-faced white man in a post-impressionist sarong,
Heading for the nearest drinking shop.
I fancied an ice-cold Guinness. A moment later
And there'd have been just me on that canvas,
Dry and wet at once, sarong slipping a bit,
Tiger a mile away and still running.

Even so, would you really see more on a larger screen,
D'you think? Or do the girls wear towels or something?

Very Clean Old Man

Do you know that land? –
Where the nocturnal tiger
Empties the rubbish bins,

Where the deathless mosquito
Stings the flesh to life
And the body runs with oils!

This splendid desk of teak
I brought from that lost land,
Large, larger than my room –

Now in this temperate clime
Cracks have run across it,
Large, larger every day.

My nails slipped through them,
Then my fingers – next
My typewriter . . .

If you change your country
Change while you are young,
Before your bones grow brittle
And your life cracks across.

These suburbs of the West,
Dog-haunted, dream
Of soap powders, and the wind
Empties the rubbish bins . . .

Maybe you should try the moon
– You cracked old man –
Once they dust it down.

The Progress of Poesy

I too would avail myself of the large and common
    benefits of modern technology.

That on the Wings of Imagination a chartered jet
    shall transport me to my inspiration.

That tapes may record the best and happiest moments
    of the happiest and best minds.

That a fine excess of surprising subject-matter
    be relayed to me by satellite.

That powerful pumps ensure the spontaneous overflow
    of powerful feelings.

That cameras shall arrest the vanishing apparitions
    which haunt the interlunations of life.

That sophisticated computers select the best words
    and collocate them in the best order.

*

A pointed stick, some vegetable dye, a strip of bark
    removed by stealth from the public park.

The Progress of Prying

It was while living in Nice
That she began the diaries called
'The Nice Diaries'.

Now that 'The Nasty Diaries'
Have been discovered
In an attic in Norfolk
We know what 'Nice' means.

We know of course what 'Nasty' means.
Although her penmanship is such
That editors have not yet established
A definitive text,
We know that 'Nasty'
Is not a lake in Finland.

Floods of light are about to be cast
Where hitherto there was no darkness.

Oyster Lament

We are a poor people, who
Cannot afford oysters any more.
The sea is a long way away
Nowadays.

Freedom we have in plenty,
Golden and tall it waves in the fields.
How lovingly somebody tilled the soil,
Manured it so richly!

But who can eat that much? What happened
To those overseas markets? Something is wrong
With the rate of exchange.

It makes such a rattle against the panes.
It is cracking the tarmac out in the streets.
They say it uses up the oxygen.

Freedom is a pearl, to be sure,
A pearl above price.
But so are oysters.
Often I think I would rather have oysters,
Their taste, their indefinable taste.

The Cauldron

'it is the time when uprightly and in pious
sober wise, naught of work is to be wrought
and art grown unpossible without the divel's
help and fires of hell under the cauldron . . .'

It had grown impossible
Very little work was getting done.
So we gave up sobriety
The other virtues as well
(We supposed that once we had them)
And we stoked the fires of hell.

How we stoked them! It was fun at first
At first there were virgins to astound
There were things to be done with things
There were things to be done in private
There were things to be done in public
At first.

It was style we believed in, not the devil
It was the devil we got.
The fires of hell are not for hire
Without the ice.
There was a saying about the devil . . .
But memory has gone, we killed it
A package deal.

Now there is no one to astound
The devil only laughs at his own jokes
Us he finds boring.
There is nothing to do in public
Nothing to do in private.
Under the dry cauldron icy fires are burning
Art is grown impossible.

Where is the future? Nothing stirring
Unless a memory stirs.
Wasn't there an ancient saying . . .
About a time when uprightly alone
In pious sober wise can work be done
And with the devil's hindrance?

The Ageing Poet

The Ageing Poet is a fertile subject for an Ageing Poet.
He knows an awful lot about it. He will relate
How the Ageing Poet's teeth are falling out (in fact
They fell out long ago, when he was a Younger Poet)
And also his hairs, as absolute as falling hopes
(An image coined in youth to point a contrary moral).
Policemen, he observes, are for ever getting younger,
Along with publishers, prime ministers and undertakers.
He may in these tolerant times remark on how a miniskirt
Or even a gymslip produces a brief but distinct convulsion
In his sluggish blood – a touchingly autumnal yearning,
Whatever the Poet himself may concurrently be up to.
(One expects to do a little better than one's fictions.)
Death too is an aspect of the human condition which may
Properly be tackled now. Mere conceit in a Young Poet,
In an Ageing Poet it testifies to courage and realism.
Also he has a number of dead friends, some quite famous,
He can mention in passing. Timor mortis is beyond reproach.

The Young and the Younger Poet may still have his teeth,
And his hairs, and his hopes that before much longer
These tedious Ageing Poets will turn into Dead ones.
He may have access to miniskirts or (come to that) gymslips,
Not to mention abortion, divorce and idealism.
But the Ageing Poet is better off. He has himself, all of him.
For a little while longer, a book or two more.

Vacillations of an Aspiring Vampire

'Guaranteed and proven,' said Rousseau,
A man of reason. 'The evidence is complete.'

I shall wear two fine pointed teeth
And a charming manner.
The ladies will be dressed in décolletage.
Nothing's more taking than necks and breasts.
The tooth is nearer to the nerve-centre,
The mouth is nearer to the heart.
This is the height of intimacy,
The rest is low and brutish.

It means a lot, when all your life
You've lacked both power and charm.
The tonics they gave me as a child
Did nothing for the circulation.
My teeth are made of chalk.
I was weaned too soon by far.
I have had trouble with bras.

'The testimony of persons of quality,'
Said Rousseau. 'Surgeons, priests and judges.'

But the laws require a tooth for a tooth.
I must be hospitable to strangers,
I shall have to give up garlic
And hide away the rood that came from Lourdes.
For first I too must die beneath the charmer's lips.

And even then, how short-lived
Immortality can be!
I see a cross nestling between full breasts,
I see myself unseen in a mirror,
I see a policeman bearing a pointed stake.

I fear those persons of quality,
Those surgeons, priests and judges,
And also the patrons of cinemas.
You could lose what little blood you have.

Buy One Now

This is a new sort of Poem,
It is Biological.
It contains a special Ingredient
(Pat. pend.) which makes it different
From other brands of poem on the market.

This new Poem does the work for you.
Just drop your mind into it
And leave it to soak
While you relax with the telly
Or go out to the pub
Or (if that is what you like)
You read a novel.

It does the work for you
While (if that is what you like)
You sleep. For it is Biological
(Pat. pend.), it penetrates
Into the darkest recesses,
It removes the understains
Which it is difficult for us
Even to speak of.

Its action is so gentle
That the most delicate mind is unharmed.
This new sort of Poem
Contains an exclusive new Ingredient
(Known only to every jackass in the trade)
And can be found in practically any magazine
You care to mention.

Or perchance not

Work – say I, coining a phrase
In the firm's money – is the
Curse of the dreaming classes.
It makes me sleep like the dead.
Where are they now, those bright
And fearful dreams I used to live?
Ingenious phantoms who conveyed
So much reality?

Dreaming will have to stop –
So says reality, aggrieved,
Wet blanket on my bed,
The tired dog in my manger.
I work, therefore for tax and
Other purposes, I am.
Then I sleep the sleep of the just,
Which is death.

Small Oratorio

Let there be pie
In the sky
When I die

In the sky
Let there be pie

There's nothing shameful in my cry!
If not for pie
What purpose in the sky?

On earth
A dearth

Whate'er his worth
That he will die
Man knows from birth

Let there be pie
Why else a sky?

A Common Interest

What is needed is a common interest,
What you might call a friendly rivalry.

My wife is patron of the cat,
I am patron of the dog.
The cat and the dog are not exactly friends,
In fact they lead a rather turbulent life.

My wife backs the cat and I support the dog.
She fattens the cat and keeps its claws in trim,
I teach the dog a trick or two.

She has equipped her client with sparklers
To brandish in its mouth.
I tell her, Peace Prizes have come from gunpowder.
We laugh. Then I teach my protégé to spit.

An atomic pellet placed in the cat's saucer?
We discuss it soberly, and we agree
To avoid any serious damage to the house.
After all, we live in it.

I think she is training the cat to drop things
From a height.
I shall fit the dog out with a helmet
And file its teeth.

How the fur flies!

We live in perfect amity, my wife and I,
Our marriage is founded on a rock.

Remembrance Sunday

The autumn leaves that strew the brooks
Lie thick as legions.
                    Only a dog limps past,
Lifting a wounded leg.
                    Was it the rocket hurt it?
Asks a child.
                    And next comes Xmas,
Reflects the mother in the silence,
When X was born or hurt or died.

Of Growing Old

They tell you of the horny carapace
Of age,
But not of thin skin growing thinner,
As if it's wearing out.

They say, when something happens
For the sixth or seventh time
It does not touch you. Yet
You find that each time's still the first.

To know more isn't to forgive more,
But to fear more, knowing more to fear.
Memory it seems is entering its prime,
Its lusty manhood. Or else

Virility of too-ripe cheese –
And there's another name for that,
One can mature excessively.
Give me cheese-tasters for psychiatrists!

Of growing old
Lots of kindly things have been reported.
Surprising that so few are true.
Is this a matter for complaint? I don't know.

PARADISE ILLUSTRATED

Paradise Illustrated: A Sequence

I

'Come!' spoke the Almighty to Adam.
'There's work to do, even in Eden.'

'I want to see what you'll call them,'
The Lord said. 'It's a good day for it.'
'And take your thumb out of your mouth,'
He added. (Adam was missing his mother.)

So they shuffled past, or they hopped,
Or they waddled. The beasts of the field
And the fowls of the air,
Pretending not to notice him.

'Speak up now,' said the Lord God briskly.
'Give each and every one the name thereof.'

'Fido,' said Adam, thinking hard,
As the animals went past him one by one,
'Bambi', 'Harpy', 'Pooh',
'Incitatus', 'Acidosis', 'Apparat',
'Krafft-Ebing', 'Indo-China', 'Schnorkel',
'Buggins', 'Bollock' –

'Bullock will do,' said the Lord God, 'I like it.
The rest are rubbish. You must try again tomorrow.'

II

'What a dream!' said Adam waking.
'I never dreamt a dream like that before.'

'You will,' remarked the Voice Divine, 'You will.'
'And worse,' He added *sotto Voce*,
Finding it hard to speak in accents mild,
Knowing what He knew.

('Why can't I ever live in the present?'
He would grumble. 'Never in the present.')

'You're luckier than I,' the Almighty said.
'I know of no one fit to shake My hand,
Let alone My equal. I'm on My own.'

'You're different,' said Adam,
'You don't need it.'
'It's wearing off,' said Adam,
'The anaesthetic.'

'Anaesthetic? What's a man like you
To do with words like that?'
He found it hard to speak in accents mild,
Knowing what He knew.

'You promised, Lord,'
Urged Adam. 'You promised me.'

'Behold her, not far off,
Flesh of your flesh, bone of your bone,'
Said He in neutral tones,
'Your madam, Adam,'
Knowing what He knew.

III

'Rich soil,' remarked the Landlord.
'Lavishly watered.' Streams to the right,
Fountains to the left.
'The rose, you observe, is without a thorn.'

176

'What's a thorn?' asked Adam.
'Something you have in your side,'
The Landlord replied.

'And since there are no seasons
All the flowers bloom all the time.'

'What's a season?' Eve inquired.
'Yours not to reason why,'
The Landlord made reply.

Odours rose from the trees,
Grapes fell from the vines,
The sand was made of gold,
The pebbles were made of pearls.

'I've never seen the like,' said Eve.
'Naturally,' the Landlord smiled.

'It's unimaginable!' sighed Adam.
'You're not obliged to imagine it,'
Snapped the Landlord. 'Yet.'

IV

If you wanted ice-cream
There was ice-cream galore
Oozing from handy ice-plants.

(But you didn't really want ice-cream:
The weather wasn't hot enough.)

If you wanted a piping hot bath
There was piping hot water
Running in convenient brooks.

(But you didn't really want a piping hot bath:
The weather wasn't cold enough.)

If you wanted petrol
There was petrol in plenty
A few inches under the Garden.

(But what was the need for petrol?
There was nowhere you wanted to go.)

If you wanted money
Money grew on trees
(But what would you do with money?).

'But Adam wants me,' Eve told herself.
Adam told himself, 'But Eve needs me.'

V

> 'About them frisking played
> All beasts of th'earth . . .'

'If we have a baby,
That elephant will have to go –
He's too unwieldy.'

'What's a baby?'
'A word I've just made up,'
Said Adam smugly.

'If we have a child,
That bear will have to go –
He's wild.'

'What's a child?'
'A word I've just made up,'
He smiled.

'If we have a brood,
That ape will have to go –
He's very rude.'

'What's a brood?'
'A word I've just made up,'
He cooed.

She said:
'But won't the wolf lie down with the lamb?'
He said:
'I think of words, therefore I am.'

VI

'Can't you let *her* name something?'
Begged Adam. 'She's always on at me
About the animals.'

'Herself a fairer flower,'
Murmured God. 'Hardly necessary,
I would say. But if it makes her happy . . . '

                        *

'What a trek!' Eve muttered.
'The animals *came* to Adam . . .
Well, Mohammed must go to the mountain.'

'What's that you said?' the Almighty asked.
But she was on her way.

                        *

'Lady's finger,' said Eve.
'Lady's smock.
Lady's slipper.
Lady's tresses . . . '

She paused.
'Adam's apple.'

'No,' said the Lord,
'Strike that out.'

'Old man's beard, then.'
She sped towards the mountain.

'Lily.
Rose.
Violet.
Daisy.
Poppy.
Amaryllis.
Eglantine.
Veronica.
Marigold.
Iris.
Marguerite.
Pansy.
Petunia.

Jasmine.
May.'

'I'm worn out,' she gasped.
'Belladonna –
And that's all for today.'

*

'She's better at names than you were,'
The Lord observed.
'They all sound womanish to me,'
Said Adam, nettled.

VII

To whom, indignantly, the Angel thus:
'Whereas by definition we are happy,
And happiness can only dwell with love –
My drift you follow, I presume?
In fact it's better than it is with you,
Since not a joint or membrane comes between.
Total we mix, we mix without restraint,
Easier than air with air, you have my word!'
       To whom thus Adam made reply:
'Too airy-fairy for the likes of me.
Myself I like a touch of flesh and blood,
Firm mounds to get my hands around.
Sex in the bed for me, not in the head:'
       To which said Raphael with contracted brow:
'If touch be what you're wild about,
Consider this: the cattle have it too.
It can't be all that marvellous . . . '
       Then 'Air with air?' asked Adam, unabashed.
'Total you mix? But show me, how –'
       'I have to go,' the Angel swiftly spoke.
So speaking, he arose. 'Tell me one thing,'
He murmured: 'Can she cook?'

VIII

'I thought I had no family ties,'
Thus the Ruined Angel sighs.

'But now I find I have a daughter.
By her I find I have a son.
I find my son has children by his mother
(Such sad dogs too).

That makes some family!
Sin and Death and all their progeny.
Not counting Lilith
Whom, truth to tell,
I can't remember well.

Eden's the right place for a family –
Where life is new and easy,
With first-grade fruits from fertile trees.
There shall they be fed
And filled immeasurably.

The present population
Consists of two small upstart squatters,
Childless, as I chance to know.
The greater good of the greater number . . .
Those two will have to go.'

IX

Satan considered the creatures.
Satan selected the serpent.

'The subtle snake, the fittest imp of fraud,'
So spoke the Fallen Angel, fond of artful sound,
So spoke the Fiend, alliteration's friend.

Unfeared and unafraid,
The silken snake lay sleeping . . .

In at his mouth the Enemy entered,
Thorough his throat commodious
Filtered our fatal Foe.

Then up he rose –
As yet the snake stood upright on his tail,
This sleek unfallen fellow,
He surged, not slithered –

181

And off the Tempter tripped,
And off swept Satan Snakeskin,
In search of silly She.

X

'All this fuss about an apple!
Now, you're a simple woman,'

Said the Snake. 'Like me.
I mean, I'm a simple serpent.

It's a conspiracy of course.
It's meant to keep you down.'

'I don't mind being down,'
Eve tittered.

'Tut-tut,' the Snake tut-tutted.
'Think of higher things!

He only wants to keep you dumb,
And that's the truth.

What's wrong with *knowing*?
Do *you* know what?'

'What's good for talking snakes
Is good for me,' she mumbled,

'Pity to leave it for the birds.'
But the Snake had scarpered.

XI

Eve chomped at the apple
(Her teeth were white and strong:
There was no such thing as decay).
The juice ran down her chin.

She ate it all
(There was no worm at the core:
There was no core).
The juice ran down her breasts.

'I have done something original,'
She told herself.
'But I mustn't be selfish.'
She plucked a second for Adam.

XII

'What about Adam?
Shall I keep it to myself,
Shall I be on top?

For who, inferior, is free?

But what if I drop dead?
He rushes off and marries someone else.

Good or bad,
Dead or alive,
On top or underneath,
I'd better share it.

I'm a simple woman.'

XIII

Sighing through all her works,
Nature gave signs of woe.
Earth trembled from her entrails,
Nature gave a second groan.

*

'What's that strange noise?' asked Eve.

'Nothing to worry about,' said Adam.
'Just cataclysms, convulsions, calamities –'

'Don't talk with your mouth full,' said Eve.

'Donner-und-Blitzen, coups-de-foudre, infernos,
Avalanches, defoliation, earthquakes, eruptions,
Tempests, turbulence, typhoons and torrents,'
Said Adam airily.

'And floods. Or do I mean droughts?'
He pondered. 'Also perhaps inclemency.'

'The Snake was right about one thing,'
Eve observed. 'It loosens the tongue.'

XIV

'Must be that Cox's orange pippin,'
Said Adam. 'I do feel queer.'

'I can see that,' said Eve.
'Come over here.'

'Here?'
'Here –

You're good at naming things.
What would you call this?'

Whisper, whisper,
Titter, titter.

'And this –
What would you call this pretty thing?'

Mumble, mumble,
Giggle, giggle.

'Hey, you never did that before!'
'It must be what they mean by sex-apple.'

'Ouch.'
'Mmm.'

'I wish there were ten forbidden trees.'
Snore.

XV

'Where are you, Adam?'

'I'm behind this tree, Lord.'

'What are you doing there?'

'Nothing, Lord. I'm naked is all.'

'Who told you you were naked?'

'I noticed it, Lord.'

'Naked, shmaked, what does it matter?'

'You can get six months for indecency, Lord.'

'What's six months to you, you're immortal.'

'It might give the animals funny thoughts.'

'My animals don't have funny thoughts.'

'The best people wear suits, Lord.'

'You were the best people, Adam.'

'The weather might change, Lord.'

'Too right. It will.'

XVI

'I should have ceased at noon
On the sixth day,'
He said to Himself.
'I think my hand was shaking.

I should have rested content
With the ounce, the libbard, the mole,
With the stag, the river-horse and the bee.

185

Would the provident emmet
Have eaten the apple?
Never.
Or the gluttonous pig?
Not ever.

The gluttonous pig
Devours his rightful truffle,
He waxes fat,
He is swollen with innocence.

But Eve – that evening's work! –
She alone would eat the apple.
And has she waxèd fat?
Lean and lecherous is she grown.

The ribs of her mate stick out
Like the ribs of a winter's tree,
Its fruits all plucked.
I should have stopped at midday.'

XVII

'The woman . . .'
The Almighty Father shook His head.
'Not like us . . . A new invention.
I blame Adam for what happened.'

'He did it out of love,'
Said the Son. 'An old invention.'

'Love was the first casualty,'
Said the Father, 'Love lies bleeding.'
Adding meaningfully,
'A lot of people are going to be in trouble
Because of love, My Son.'

Then thick as stars
A solemn look on every face
The angels drifted in,
Ending this little tête-à-tête.

XVIII

'Son of My bosom,
My Word, My Wisdom, and My Might!'

'Too easy, loving me.
In me You love Yourself,
Your Word, Your Wisdom, and Your Might.'

'My sole complacence,
Radiant image of My Glory!'

'What I mean precisely.
Much further, Father, You must love –
And love what's hard to love.'

'Too much talk of love.
Die man, or someone else must die.'

'Account me man *pro tem.*
Pro tem account me man.'

Nothing was said about a cross.
By now the quire was in full swing.

XIX

So the Archangel, out of pity,
Now disclosed the liberal arts
That should relieve man's fallen lot.
'Like music, painting, plays and books.'

'Long books?' asked doleful Adam,
Whom the stern Angel had apprised
Of death and rape and guns and hunger.

'A book there'll be,' the Angel said,
'About this very business –
A poem of ten thousand lines, which one
Called Milton shall compose in time to come.'

'Oh dear!' Then Adam brightened.
'Am I the hero of this book perchance?'
'Not quite the hero,' Michael mildly said,
'And yet you feature largely in it –
God, not unnaturally, is the hero.'
'Should have known,' groaned Adam.

'Although there are – or will be – those
Who claim the hero really is – or will be –
Satan. As I of late foretold,
Henceforth the human race is fallible.'

'That circus snake?' hissed Adam scornfully.
Eve hid her blushes in her work,
A garment she was knitting, made with
Real lamb's wool, tight-fitting.

'In my opinion, which I trust
You won't repeat,' the Angel whispered,
'The hero really is the Son,
Called Jesus, even though his lines
Are fewer in the poem than are mine.'

'And me?' Eve raised her eyes. 'Am I in this –
This book of yours? Or, as I well suppose,
Are all the characters men?'

'Indeed you are!' the genial Angel cried,
'Without an Eve there'd be no tale.
While Mr Milton's not a woman's man,
He does your beauty justice, and your brains.'

'A female intellectual?' Eve grew vexed,
Old-fashioned in her ways as yet.
'No,' spoke the nervous Angel, blushing more,
'I only meant, not just a pretty face.'

Eve held the knitting to her breast.
'By me the Promised Seed shall all restore.'
And Michael knew the time was ripe to leave.
'All – or some,' he murmured at the door.

XX

There was the escort, in position,
Presenting fiery arms and flaming brands –
Handpicked Cherubim, four faces
To each man and pairs of eyes to match.

'A splendid turn-out!' Eve declared,
Waving graciously.
'Not like our coming in,' said Adam,
Grabbing her wrist.

'Hurry up please!' said the Archangel.
His men were sweating, the Sword of God
Had started the stubble burning.
'We don't want any trouble, do we?'

XXI

'I had no voice,'
Sighed Adam. 'No real choice.'

Eve wiped a tear and said,
'I wanted you, alive or dead.'

'Whether boon or curse.'
'For better or for worse.'

'Come fair or foul, we have to eat' –
She watched for herbs beneath their feet.

The world lay there, and they could choose.
He said, 'I'll learn to make you shoes.'

XXII

'Why didn't we think of clothes before?'
Asked Adam,
Removing Eve's.

'Why did we ever think of clothes?'
Asked Eve,
Laundering Adam's.

XXIII

'What was she like?'

'Who?' asked Adam warily.
'What was who like?'

'Come off it, Adam,
Don't act dumb.
Lilith, of course.
Who else?
Yes, come to that, *who* else?'

'Lilith? Lilith, you say?
Lilith . . . Lilith . . .'

'The one you had before me.
You haven't forgotten already?
The screech-owl –
Was she too much for you?'

'Before you? Fat chance I had
Of having anyone.
Count my ribs if you like.'

'It's common knowledge.
No use your denying it.'

'Oh *Lilith*! – I remember –
The one who shacked up with your fancy man
The Snake.'

'My *what*?
I only met that Snake once in my life.
We only talked.'

'So you say.
As a matter of fact
There's no such woman as Lilith.
The Lord God told me so.'

'First she hangs around with Satan,
Then you tell me she doesn't exist!'

'It's all made up.
It's an old wives' tale.
It's a Jewish conspiracy.'

'So what are you?'

XXIV

'I gave him a very nice name,'
    said petulant Adam.
'I called him SNAKE.
I even gave him a second name,
I called him SERPENT.
There's no gratitude in the world,
They bite the hand that feeds them.'

'The tiger gave me a nasty look,'
    Eve got in quickly.
'The elephant made a vulgar noise,
The cat has scratched my nose,
I found a cockroach in the kitchen.
I'm glad it wasn't me
Who named them!'

XXV

'Death,' said Adam in funereal tones.
'That's the worst of what we've done.
As for the rest of it – that's life.
But Death's a killer.'

'Death?' said Eve, propped on her mop.
'Death would have been deadly in Eden,
Most unsuitable.
But we're not in Eden now.
Other places, other manners.'

'You have a point,' said Adam,
Gladdened. 'It might be held
That, if not exactly vital, Death
Is at all events not wholly inadvisable
In the conditions now prevailing.'

'How well you put it!' Eve admired.
'And may I now remind you
It's time you thinned the carrots out
And there's a lamb to slaughter.'

XXVI

'What's all this
About a sense of sin?
It's rot!' cried young Cain,
'It's the opium of the people.'

'It's all we have, son,'
Said Adam. 'We grew up with it.'
'What's opium?'
Asked Eve suspiciously.

Later Cain killed his brother
For business reasons.
'You see?' his parents said,
'We told you so.'

XXVII

'I told Him flat.
I said: Look here, the ape can't speak to the fish,
The bird can hardly converse with the cat –
How on earth am I to talk to any of them?

I am here, He said, in that way of His. –
Yes, I said, but you'll soon be off.
The country's all right, it's full of nature,
But is it enough?

Happy? He said. You mean you want to be *happy*? –
Well, it is my birthright, isn't it?
I could see He was getting ratty.

Very well, He said, fair's fair.
What next I bring shall please thee, I'm no liar. –
An odd expression came over His face –
Thy wish exactly to thy heart's desire . . .'

*

Then Eve came home from the dentist.
'You're always talking to yourself,
Adam! You should see a psychiatrist.'

XXVIII

Enoch died untimely, he was only 365.
There was bad blood between Adah and Zillah.
Lamech killed a youth in a brawl.
Certain angels got physical.
There was wickedness in tents.
There were giants in the earth.

                    *

'What about that Promised Seed
You promised?' Adam asked.
'There's a lot needs restoring.'

'Not yet . . . I have a feeling,'
Said Eve. 'Around the year 1000.'

'Hmm,' said Adam. 'Even so . . . '
'It's not something one can rush,'
Said Eve. 'At your age, too!'

'Even so?' Adam asked nicely.
'I have a feeling too.'

XXIX

There were giants in the earth.
Some said they were angels
Lusting after pretty Jewish girls.
Others said they were scholars
(That is, 'mental giants')
Lusting after pretty Jewish girls.

Either way, their offspring turned out ill,
They went to the Devil.
Who had been living in partial retirement,
Teaching himself German:

'Ich bin der Geist der stets verneint! –
Oh very well then, I accept.'

'I knew this would happen,' said God.

XXX

Jabal had erected the tents,
Jubal was working the hurdy-gurdy,
Tubalcain was running the Ghost Train.

There were crowds of people.
Seth and Enos, Irad and Lamech, Adah and Zillah,
And the sons and daughters of sons and daughters.

There were crowds of people,
There was candy floss and Brighton rock.
It was all go.

'We are making on the swings,'
Said Adam. He was 900 years of age.
Eve counted the takings in silence.
She was slightly younger.

'I'll take a quick look,'
Said Adam.
Coming back, he told her,
'We are losing on the roundabouts.'

XXXI

'Cities were all right when they were towns,'
Adam grumbled.
'Now it's all rush and noise and pollution.'

'Only this morning I stepped in something
Nasty,' said Eve. 'Left by a dog.
It shouldn't be permitted.'

'Don't blame me then,' said Adam.
'Dogs were always like that.'
Give a dog a bad name, he thought.

'And horses are no better. Look –
Just in front of the house!
You'd wonder where it all comes from.'

Adam rushed for a shovel.
'Can't have you stepping in it, dear.'
It was just the thing for his allotment.

XXXII

The days of Adam were 930 years.
He sat in the market-place
With other senior citizens.

With Seth, who was just 800,
Enos, who was 695,
And Methuselah, only 243.

'They're not the men
Their great-grandfathers were,'
Said Seth.

'Lamech's kid Noah cries all night,'
Said Enos.
'Howls when they bath it.'

Said the youthful Methuselah:
'They've all been spoilt.
I blame their mothers.'

'It was different in my day,'
Said Adam.
'People lived for ever then.'

XXXIII

Adam was old bones, lost in his clothes.

Eve thought of how they met. Long,
Long ago. It must have been on holiday.
Warm the air, and full of scents.
And he was wearing nothing, not a stitch.
She blushed a little, pleasurably.

Adam's dreams were not so pleasing.
He woke, delirious and sweating,
Tearing at his twisted nightshirt.

'Eve . . . my side is bleeding . . .
A spear stuck in my side. My rib is hurting.'
'Look,' she said, 'there's nothing there,
No sign at all, dear. Go to sleep.'

He fell back, mumbling indistinctly
About a seed, a promised seed.
She wiped his brow, and smiled:
At heart he'd always been a gardener.

XXXIV

'The Lord called Adam *Man*.
But Adam called me *Life* itself.

Adam is dead now, man is not.
The Man is dead, but long live man.
For Life goes on.

Priests everywhere, police
And politicians, laying down the law.'
The old girl grinned.

'And I it was who broke it . . .
Poor Adam, poor old Man,
Our sin was self-sufficiency he said.'

Eve did the crossword by her lonely fire.
'That which clasps, a tendril: RIB.'
She died soon after Adam.

196

The Verb 'To Think'

The verb 'to think'
Is represented by the same sign as
'To yearn for', 'to be sad',
'To be unable to forget'

In many cases
'To love' is rendered as 'to like'
(Thus, a man is said to 'like' his wife)

The symbol for 'feeling' also signifies
'A certificate of merit'.
It is meritorious to feel,
Though 'to love' is more often 'to like'

For love comes in two forms alone:
That of a mother for her child
And carnal

The verb 'to do' is homophonous with
The verb 'to pick someone's pocket'.
This, we should note, is not
(As it is in some languages) slang

Slang would be to say that you 'love'
Someone or something when more properly
You 'like'

The verb 'to be'
Seems not to exist.
Yet you cannot say that the verb 'to be'
Does not seem 'to be'

'To be sad' does not imply 'to be',
It implies 'to think'.
'To think' is 'to yearn for', 'to be without',
And to be a man of feeling is to be certified.
'To like' is more common than 'to love'
And 'to be' is most uncommon –

Unwise it is, in the long watches,
When the leaves litter the misty streets,
To be reading a Japanese lexicon
And to be unable to forget.

Old Colony, New Country

More put inside!
Why, the streets must be quite empty
That I once knew.
But no, there's plenty left to walk to work,
Just a bar or two with tables free.

Some of them in for the second time,
Accident prone.
The third time's time for others:
No matter who might make the rules,
They broke them. Mindless mules –
Or else retired to finish off a thesis?

One I knew convinced the British,
An intellectual he, no peasant,
That he should keep his shoes.
Long since that empire passed away,
I wonder what he wears today.

There's always one snake in the grass.
And God gets mad. Then a second snake.
And God gets madder.

Such a burden to the State
And irreproachables who pay their tax,
To keep these old black sheep in clover!
And shoes moreover.

That was why the juries were disbenched,
Squeamish teams who flinched from death –
A man can eat a lot of free food in a life,
Life sentence.

No, the streets will still be busy,
The restaurants too
That I once knew –
The economy's in clover.
'The value of a thing
Is reckoned by the life exchanged for it.'
Gaols are no enormous luxury.

Many Voices

And we chatter ourselves to death
In many voices
We do God in many voices

In the magazines, on the air
We mouth towards extinction
In many crises

We do Death and Disaster
In many voices
And we sit at home and drink the proceeds
In many doses

We make our own ways to death
In lower cases
Slowly or quickly
By various vices

Whereas in other places
Over the virtues
Under the vices
Flows a large and fruitful silence

We do silence too in many voices
Many silences speak for survival.

Lives of the Gaolers

Let us recall Mr Dagg, whatever his lineage,
He showed himself a tender turnkey.

There was a poet, a Mr Savage, who sometimes
Lived up to his name, a troublesome bastard
Of the Earl of Here and the Countess of There.
His illegitimate father died of a distemper:
'I can only,' he said, 'be a godfather to you.'
His illtempered mother desired to send him
To the American plantations in secret.

For a debt of eight pounds, owed to a Mrs Read
Who managed a coffee-house, the poet was consigned
To Newgate and the care of Mr Dagg, the keeper.
This happened on his 45th birthday, for Mr Savage
Was proficient in misfortune.

Inspired perhaps by his mother, he had once written
A play, *Woman's a Riddle*, also a poem about Bath,
The Triumph of Health and Mirth. They availed him
But little. 'But now,' he said, 'Freely I sing
In my cage, not a gaolbird but a bird of the Muses.'

'Here,' said Mr Savage, 'I have a room to myself.'
Moreover he fed at the keeper's own table,
And often was taken for walks in the meadows.
Never in his life had the poet had it so good.
In his death he was buried at Mr Dagg's expense.

Modern scholars contend that in fact Mr Savage
Was of low and legitimate birth. There is a book
On him, grudgingly titled *The Artificial Bastard*.
But the moral of the story remains unaffected:
Virtue, as Dr Johnson observed, is most laudable
In conditions which make it most difficult –
Like being a gaoler. Mr Dagg should not be forgotten.

The Roosters Cry

'"Deny deny deny,"
is not all the roosters cry'
– Elizabeth Bishop

Quite true. In fact we find
that what they say is of this kind:

'Good morning, good morning, good morning
– which was just to clear the throat,
not to gain your attention,
the few of you that are out and about
this good morning.
Though according to the sheikhs who know,
Doomsday comes
when once I fail to crow,
and not good morning.

I propose to spend the rest of the day
in glaring, as is natural to my eyes,
in asserting my chest (by nature assertive)
and my two iridescent legs, while my bill
(the which is black and as the jet it shines)
shall be incisive.

I shall feed with an easy grace,
pausing only to push some lesser breed
out of the way.
Hen is cock's joy, so we say:
I shall couple with precision and dash
and to such effect as will put you to shame
should you be watching
(which is not why I do it).
I am not called what I am for nothing.

The cock who crows three times
is either out of breath or wasting it.
If someone in the vicinity
happens to be engaged in betrayal
whether positively or negatively,
it is purely coincidental.

I do not know the word "betray".
Nor the word "strut" – which seems to mean
"to tread" or even "be",
it being necessary to pass from one grain
to another, one hen to the next, in a manner
as sure-footed as can be. Hence
my excellent legs, et cetera.

Denial is equally foreign.
Even the oldest rooster in the yard
– his legs have gone, his feathers gone,
his comb has gone, et cetera –
he doesn't cry "Deny". What he says,
if anything (his voice has gone),
sounds rather like "Accept", or "Take".'

'Good morning' just once more,
and then he's off to make cock-sure.

The Answer

It was with aesthetic satisfaction and
Intense moral relief that he began at last
To see a figure forming in the carpet.

'Excuse me sir,' the carpet murmured faintly,
'I am gratified that you perceive a pattern,
But people continue to walk all over me.'

And there it was, the figure in the carpet!

The Word

The sage said: We are all books
In the great Library of God.
(He was a bookish person.)

One asked: Does He ever
Take us out?
We spend our years as a tale that is told.

The sage said: His will be done
In the Library as it is Elsewhere.

One asked: But perhaps
He is only interested in first editions,
Not in reprints, abridgements, strip cartoon
Or other adaptations?

The sage said: His love speaks volumes.
He is a speed reader. He is no respecter
Of Bestseller lists.
He suffers the little magazines to come unto Him.

Some hoped their jackets would be clean
And well pressed when the call was heard,
Their loins girded about and their lights burning.

God thought: I wrote all the books,
Now they expect Me to read them.

Matter of Opinion

And the Lord God was grieved at His heart
And said:
'Opinionation is upon the face of the earth.
My thoughts turn towards water.

For it is taught in the schools
That the people shall have each their Opinion,
And my Opinion is as good as your Opinion,
And if your Opinion is not as good as my Opinion
Nevertheless it is second-best.

To offend an Opinion is more shameful
Than to see thy father's nakedness
(He should have clothed himself in Opinions).

The people have rewritten My Commandments –
Thou shalt have no other Opinions before thine own.
Thou shalt honour and obey thy Opinion.
Thou shalt not covet thy neighbour's Opinion,
        nor his ass's,
Thou shalt not steal it,
        nor bear false witness against it,
        nor commit adultery with it.
Blessed are the Opinionmakers, for they shall be called
The children of God –

And what, you may ask,'
Thus spake the Lord, 'is My Opinion?
For I also have a right to My Opinion . . .
My Opinion is that things will change
When the people have nothing to eat but their words.'

Seaside Sensation

The strains of an elastic band
Waft softly o'er the sandy strand.
The maestro stretches out his hands
To bless the bandiest of bands.

Their instruments are big and heavy –
A glockenspiel for spieling Glock,
A handsome bandsome cuckoo clock
For use in Strauss (Johann not Lévi),

Deep-throated timpani in rows
For symphonies by Berlioz,
And lutes and flutes and concertinas,
Serpents, shawms and ocarinas.

The sun is shining, there are miles
Of peeling skin and healing smiles.
Also water which is doing
What it ought to, fro- and to-ing.

But can the band the bandstand stand?
Or can the bandstand stand the band?
The sand, the sand, it cannot stand
The strain of bandstand and a band!

Now swallowed up are band and stand
And smiling faces black and tanned.
The sand was quick and they were slow.
You hear them playing on below.

Poem Translated from a Foreign Language

My mother once told me
Never to sleep with flowers in the room.
Since then I have never slept
With flowers.

I have slept with prostitutes
(But never with flowers in the room).
In general they were amiable
A few were not unhumorous
One was rather acid.

None of them smothered me.
One of them desired a silver pen
Another removed a traveller's cheque
But never my precious oxygen.

My mother never told me
Not to sleep with prostitutes in the room
She did not tell me to stay awake.
These days I sleep with pills
I sleep with daffodils in the room.

Anthropos feels his Age

Anthropos has lived a long time
And many lives. Chieftain and pedant,
Merchant and peasant. Has borne children,
And arms, and filled long cemeteries.

Anthropos feels weary, though he can ride
In a car, a tractor will pull his plough,
And machines add up and take away.
They also multiply.

He has fewer things to hide from.
The sabre-toothed tiger has left,
There are laws against drunken driving,
The Black Death has been abolished,
Religions are not supposed to burn him.

But there are fewer places to hide.
Once there were torrents to cross,
Forests to explore, and the nature of God.
The objects that squat on his desk
Afford him no refuge.

Television shows him places to hide
Where he cannot lose himself.
He is versed in the anatomy of man
And the biology of woman. The world
Is full of good intentions and dislike.

Anthropos looks at the moon –
It is pretty from a distance.
The sabre-toothed tiger is not found there.
Nor thrilling and dreadful women. Nor gods.
Machines love it.

Anthropos feels he has lived too long,
Or else not long enough.
He can turn petulant at any moment
And throw things about.
Then he will have to put them back.

History of World Languages

They spoke the loveliest of languages.
Their tongues entwined in Persian, ran
And fused. Words kissed, a phrase embraced,
Verbs conjugated sweetly. Verse began.
So Eve and Adam lapped each other up
The livelong day, the lyric night.

Of all known tongues most suasive
Was the Snake's. His oratory was Arabic,
Whose simile and rhetoric seduced her
('Sovran of creatures, universal dame').
So potent its appeal –
The apple asked for eating,
To eat it she was game.

Now Gabriel turned up, the scholars say,
Shouting in Turkish. Harsh and menacing,
But late. And sounds like swords were swung.
Fault was underlined, and crime defined.
The gate slammed with the clangour of his tongue.

Eden was gone. A lot of other things
Were won. Or done. Or suffered.
Thorns and thistles, dust and dearth.
The words were all before them, which to choose.
Their tongues now turned to English,
With its colonies of twangs.
And they were down to earth.

A Child's Guide to Welfare

The babe insures the teat
The hope insures the ray
The pot insures the teatray

The driver insures the car
The mistress insures the pet
The housewife insures the carpet

The cat insures its purr
The model insures her pose
The planner insures his purpose

The college insures its don
The composer insures his key
The cart insures the donkey

The glutton insures his tum
The cow insures her bull
The rough insures the tumble

The slaver insures his dhow
The lawyer insures the *re*
The bride insures her dowry

The bank insures its cod
The master insures his piece
The lover insures his codpiece

The merchant insures his wreck
The hangman insures the cord
The runner insures his record

The hand insures the arm
The I insures the Me
The state insures the army

The consumptive insures his cough
The shark insures its fin
The corpse insures its coffin.

A FAUST BOOK

'Le personnage de Faust et celui de son affreux compère ont
droit à toutes les réincarnations . . . J'ai donc osé m'en servir'
– Valéry

'Everyone should write a *Faust* of his own' – attributed to Heine

'Euch ist bekannt, was wir bedürfen,
Wir wollen stark Getränke schlürfen . . . ' – Goethe

I

Dr Faust discourses on the subject of training and heredity

Out for a healthy walk, Faustus picks
Up a poodle, or vice versa.
A comical little fellow, it chases sticks,
Crawls on its belly, races in circles.

'Perhaps you could train it,' says Wagner.
'No doubt it would respond better
Than many of my students,'
Remarks the Professor.

'Even so,' his assistant ventures,
'There's the question of pedigree.'
'A dog's a dog for all that,'
Says the Professor unctuously.

'To heel, Blackie!' Blackie wags its tail.
'Man's best friend,' says the Professor,
A shade sentimental, it being Easter.
The dog assumes a toothy smile.

Faust in his study

'The interesting questions go unanswered –
Undue suffering, undeserved success . . .

Divinity talks all the time of better times to come,
Science invents false limbs and true explosives,
And Alchemy turns lead to lead at vast expense;
Art thrills with piggish gods and puffy goddesses,
While Law invokes the laws its lawyers legalize,
And Logic says it all depends on what you mean . . .

So must I live and die in Aristotle's works?'

*

'Then what is life,'
Faust paced between his piles of parchments,
'Except a slower form of suicide?'

A first-class speaker,
He found himself a first-class listener too.

'This is indeed an honour,'
Yapped the dog. 'I could crouch here
Listening to you for all eternity,
I mean all day.'

'I thought there was more to you
Than met the eye,' said Faust,
Patting him.

'Talking of suicide,' the dog remarked,
'Perhaps I can be of some small service,'
Patting Faust.

Mephistopheles enlightens Dr Faust

'Why, if I may ask, did you appear
In the shape of a dog, Mephistopheles?'

'As you will know, our liturgies
Run widdershins, dear Doctor–
A backwards Paternoster,
Commending Cain, rejoicing in Judas,
The Black Mass et cetera.
In my view, often a little vulgar –
Part of the hearts and minds manoeuvre
To win over the masses . . .

So in our parlance the Lord God
Is the Drol Dog,
Ergo, a poodle.
Nothing could be more simple . . .

Talking of which, every god has its day –
What would you like to do with yours?'

Faust prefers to succeed through his own efforts, but Mephistopheles warns him of the obstacles

'You slog away at it, work, work, work,
Without pulling any strings –
Because your pride won't let you,
Or there aren't any strings to pull.
And at long last you get There –

Looking like the sole runner
In a marathon – first or last?
Your tongue hanging out,
And good for nothing. You made it,
And it unmade you.

And the assembled spectators
Will say, all three of them:
Hurrah, this old fellow got Here
Under his very own steam!

(You are leaking from every piston.)
They move away, and Here goes with them.

We all need someone's help –
So why not mine?'

Mephistopheles desires a few lines in writing

'Being of sound mind (and all that mumbo-jumbo), I (name in full)
by these presents (already?) do covenant and grant, in consideration
of 24 (Twenty-Four) years of service by Mephistopheles, the which
(which?) shall bring me anything, or do for me whatsoever, that I
give them (names of Lucifer and self) full power to do with me at
their pleasure, to rule, to send, fetch, or carry me (that should cover
everything) or mine, be it either body, soul, flesh, blood, or goods
(curious expression), into their habitation, be it wheresoever (how
discreet!) . . .'

*

'Here are the papers,
All ready for signature –
I'm yours for 24 years,
You're mine thereafter . . .

Now where's my penknife?
Quick, before it dries . . .
Blood is thicker than ink,
And I hate to strain my eyes.

I've immense respect for symbols,
They mean a lot to me.
Noughts and crosses, for example,
And the loaden tree.'

(Some other words were heard in Faustus' mind.
There is in love a sweetnesse readie penn'd:
Copie out onely that, and save expense –
But reason could not tell him what they meant.)

'Anaemic forsooth!
Lucky I'm not that young Count
With a sharp sweet tooth
And a taste for throats . . .

Words appearing on your palm?
Sounds most unlikely!
It looks like what they call a Rorschach Blot,
They say it shows your personality.

That's right! –
Now he who runs may read . . .
Oh do stop fussing!
A little spittle clears you of this deed.'

The Professor reminisces

'The likeliest pupil I've had for ages
Was a young fellow called Hamlet,
Foreign but well connected . . .

Not much of a drinker,
But popular with the student body,
And also with the lasses.

An ingenious mind, though always tardy
With his essays . . .
I don't see him as a seminal thinker,

More as a clever *vulgarisateur* . . .
Given the right circumstances,
He ought to go far.'

Mephistopheles declares himself a man of the people

'Myself I favour democracy –
Though as a Prince's man
I can't admit to it openly . . .

It's only right that man should choose freely –
Enjoy pornography of his own accord,
Not be dragged to it like a hijack victim.

Or even make the porn himself
For his own good reasons, like art or money . . .

213

It's the free agent who interests us,
The rapist – not the raped,
Not normally . . .

Surely the power of election must rest
With the people – of whom there are many.
The process may be time-consuming,
But in the end it eats up . . . more than time.

I was raised among powers and principalities,
But deep in my heart I'm a democrat.
I believe there's a soul in everybody.'

Faust rebukes Mephistopheles
for his vulgar use of language

'Knowledge is my bread and butter,'
Faust declared, 'knowledge and truth.'

'Quite,' concurred his affable companion,
Helping himself to more venison.

'The pursuit of knowledge,' said Faust,
Swallowing fast, 'is beset by perils . . .'

'A price on your pate as it were?'
Mephistopheles suggested.

'A manacle on the mind,'
His host corrected coolly.

'I think I follow you –
Adam and his madam paid a price.'

Faust reached for the fruit.
'Precisely,' he said indistinctly.

'They lost a garden,' said his guest,
'But they found a world.'

'At times,' Faust spoke sharply,
'You talk just like a journalist!'

Wagner sniggered. 'Your famulus
Grows familiar!' Mephisto hissed.

Faust grows impatient with his companion's dark mutterings

'You're forever talking rot,' snapped Faust.
'*Which way you fly is hell*, and suchlike . . . '

'And what do you intend,' he groused,
'By *this is hell and you are in it*? –

This is the University Senior Common Room,
And you are in it as my guest!'

'I do repeat myself, I fear,'
His shamefaced friend confessed.

Mephistopheles introduces Faust to the fourth estate

Under the good solid name of Gutenberg Inc.
Mephisto instructs Faust in the art of printing . . .

(The wicked sprites strike up on cue:
      'We have done our best or worst,
      Faust or Fust shall be the first –
      Black letters on white background seen
      Will prove our powder-magazine . . . '
As infantile as ever.)

Whereupon Herr Gutenberg declares himself bankrupt
– 'This fairy gold does tend to depreciate' –
And hands the business over to Dr Faust or Fust.

'I have taught you everything I know,
The rest is up to you, my boy –
Take care of your running head, and
Your tailpiece will take care of itself.'

'There aren't many authors about, ' Faust complains.
'The Imperial Treasury ought to make grants to writers.'

'I know an author,' says helpful Mephistopheles,
'Whose work is out of copyright moreover.'

*

215

So Faustus printed the Bible,
In black letters on white background.
And after the copies were all bound
He discovered a word was missing –
'Thou shalt commit adultery.'

'You've turned the Good Book into a naughty one,'
Mephisto laughed. 'I can teach you nothing.'
'You can't get decent proof-readers,'
Moaned Faust, building a bonfire.

He began again, and when the books were distributed
Someone detected a dangerous heresy –
'The fool hath said in his heart there is a God.'
Faust was fined £3000 and the book was suppressed.

'Here's fool's gold to pay a fool's fine,'
Mephisto offered generously.
'You had better specialize in French romances.'

Faust requires Mephistopheles to describe hell and heaven for him

'Tell me, I command you,
What is that place called hell?'

'Hell is grey and has no bottom –
Which reminds me of a joke I heard . . .'

'It was of hell I asked to hear,
Not your stale jokes.'

'But hell is of no substance,
A confused and hence confusing thing –'

'It is yourself that you describe,
I think.'

'So be it – hell is,
Well, it's empty.
Oh there are crowds of people all around.
But hell, you feel, is – empty.
The word is emptiness.'

                    *

'Now let us speak of heaven.
What manner of place is that?'

'I lack authority to speak of such.'
(Though not to miss it.)

'Yet you were once in heaven,
Or so I've heard.'

'Some time ago, it may have changed . . .
But – heaven is full.
That's not to say it's chock-a-block with people.
Simply, full – that's what you feel.
The word is fullness.'

'For one renowned for eloquence
I find you sadly tongue-tied!'

'Whereof one cannot bear to speak,
Thereof one says but little.'

Faust asks for trouble

'Now tell me, Mephistopheles – who was it made the world?'
That was a red rag, that would put the cat among the pigeons.
Faustus was flushed with wine, he was feeling foolhardy.

'Ah,' said the other urbanely, 'an interesting question.
There were several of us there at the time, and you know
What these collaborative efforts are. Team-work's the answer.'

217

Faust has doubts about the bargain

'In return for my immortal soul
I looked for something grander –'
So sighed Faustus, full of dole
' – Than roosters laying addled eggs,
Sour grapes out of season,
And flat beer drawn from table legs.'

Mephisto couldn't say so,
But the going rate for used souls,
Even immortal, was rather low.

'You polished off the grapes for sure,'
He answered drily.
'And don't forget, I helped Old Moore
Predict the grave indisposition
Of a member of the Royal Family –
Likewise Mother Shipton
And the dissolution of the monasteries . . .'

'Poor devil, he's slipped a cog,'
Faust cried jeeringly.
'The monasteries are built on solid rock
And the Kaiser's in his prime!'

'When you've been in hell as long as me,'
His friend wailed piteously,
'You'll find it does things to your sense of time.'

Faust journeys to unknown regions

Dolphins bore Faust to the bottom of the sea.
Dragons swept him into deepest space.

The dolphins carried him back to land.
The dragons brought him down to earth.

'I have been where no man was before!' he cried.
'To be is not to conquer,' sniffed Mephisto.

*

'And I saw a new heaven and a new earth,'
Faust said. 'For the first heaven and the first earth
   were passed away.'

'I don't think so,' said Mephistopheles, seized
   with unaccountable mirth.
'Not in what we might loosely term your day.'

*

'When you are rested,' Mephisto promised,
'I shall take you to the sage called Freud.
He will guide you on another journey,
He will charm your sins away.'

The Doctor displays his erudition in philology

'Do you mind if we go by broomstick?'
Mephistopheles asked.
'It wouldn't look right, arriving by cab.'

*

'The word *sabbat* or *sabbath*,' said Faust,
'Does not derive from the mystic number *seven*,
As is commonly believed . . .'

'Number seven – go to heaven,'
Mephisto murmured, busy at the controls.

'An amusing though scarcely tenable theory
Has it that the word comes from *Sabazius*,
A Phrygian deity akin to Bacchus . . .'

'Quite a strong head wind,' the pilot grumbled.
'Steeds would have been steadier.'

'But the consensus of opinion has it that *sabbath*
Comes from the Hebrew *shābath*, meaning *to rest* . . .'

'There's no rest for the wicked.'
The pilot corrected a nose dive.
'New brooms don't always sweep clean.'

'My own view, necessarily tentative,
Is that it stems from *saba*, as in *sabaism*
Or *star worship*, properly *host of heaven* . . .'

'In the vernacular,' Mephisto interjected,
'*Army of devils* . . . Bend your knees, please,
We're coming in to land!'

'Interesting word, *knee*: from Germanic *knewam*,
Cognate with Latin *genu*, whence quite possibly
We take *ingenuous*, inclined to kneel –'

*

'I can't see you enjoying the orgy,'
Mephisto scolded, 'with a broken ankle.
For you the word is *shābath* . . .'

Faust asks another hard question

'I charge you, Mephistopheles,
Tell me this –
Why is it little children suffer,
Guiltless beyond dispute?'

'It passes understanding,'
            came the pious answer.
'It may surprise you, but in hell
We need to keep child-murderers and molesters
Segregated from the rest. Feelings run high.

No, you can't lay that disgrace on me!
I would rather a millstone were hanged around my neck
And I was drowned in the depth of the sea,
For example.

Even if I'm fishing in the river
And catch a little one, I always throw it back.'

Faust is struck by fears for his career

'There's an academic joke,
A funny story, well, meant to be funny . . . '

'Indeed?' says Mephistopheles glumly,
Not very fond of Faustus' funny stories.

'Well, you see,
A soldier squatting at the foot of the Cross
Says to another:
If the fellow's as bright as you say,
Excellent speaker, always helping people . . .
Then why is he *up there*? – '

Mephistopheles starts laughing.
'Not yet,' says Faust testily,
'I haven't got to the point yet . . .

Then the other soldier says:
Ah, no publications!'

'They came later,' Mephistopheles remarks,
Forgetting to laugh.

'It's a *joke*!' Faust snarls,
'About getting on in the world . . . '

'I thought you were what they call
Permanent,' his friend says soothingly.

'But Wittenberg isn't an awfully prestigious
Place . . . I was thinking about Berlin,
The Sorbonne, or even East Anglia . . .

I might produce a book on Metaphysical Evil,
Of course – with some help from you . . . '

'By all means,' his friend replies.
'To coin a phrase – be damned and publish!'

Mephistopheles reveals the marvels of science

This was one of the scientific marvels
Which Mephistopheles disclosed to Faust.

'A sound physicist is a demi-dog,' said he.
'Ignei, aerii, aquatici, terreni spiritus salvete!'

Whereupon appeared a vision of muscular demons,
Busy with choppers and chisels in a fiery cellar.

'What are they doing there?' Faust asked.

'They are engaged in splitting the atom.'

'Why should they wish to do such a thing?'

'Why? Because it is theologically impossible,
Because it annoys the Almighty,
And because it makes a devilish big bang –
That's why! . . .

On kai me on!'
He dismissed the sweating spirits.
'Myself, I cannot endure loud noises.'

Faust quizzes Mephistopheles about love

'But what can fallen angels know of love?'

'We know desire that nothing will relieve.'

'There surely must be whores enough in hell!'

'And all retired . . . My theme is something else.'

'What do you do with love instead of loving?'

'We sublimate our fruitless longing.'

'So hell is paved with high ideals . . . But how?'

'We take an interest in such as you.'

Mephistopheles' minions ransack the world's libraries for Faust's gratification

Mephisto plied Faust with rare books
And curious incunabula.

Such as Aretino's *Sonetti Lussuriosi*,
The *Kama Sutra*, with helpful diagrams,
Homer on Ulysses (a case of mistaken identity),
And some obscure poems about flowers by a Frenchman.

Also *The 120 Days of Sodom*, abridged for busy readers,
A guide to ill-starred taverns called *Lolita*,
The Story of O by Oh!
And a checklist of forthcoming titles from the Olympia Press.

(Later, he knew, he would have to do much better,
To less effect.
Leisure would pose a serious problem.)

'Why all this sex stuff?' asked Faust scornfully,
His bed heaped with it.

'You should show more respect.'
Mephisto sounded hurt. 'It was the very first way
Of transmitting original sin . . .'

Mephistopheles plays the go-between

'What is love?'
    Mephisto asked Faust cheerily.
''Tis not hereafter.' (None there embrace.)

'A boy-friend is a girl's best friend,'
    he then told Gretchen earnestly.
'And diamonds next.' (Paste in her case.)

These things were quick to fix.
Faust lived in a big house with lots of space,
Gretchen's cottage was dark and poky.
The sexes lent themselves to sex.
He hadn't really needed jewels. Or poetry.

Faust and Gretchen walk in the garden

'No, I don't see much of the priest –
Actually I'm a Doctor of Divinity myself . . .'

If Divinity was sick
He was the man to cure it.

'Do I believe in God, you ask.
You're a natural philosopher!
Can anyone say he believes in God?
What is meant by *God*?
What is meant by *believe*?
What is meant by *I*?
Can we employ these words any more?'

Gretchen felt that she could,
Though in fact she rarely did.

'But when I speak of *hand*
I know what I mean, I can touch it . . .
Ah, let me kiss it!
What do you mean, I don't know
Where it's been?'

It had been in soap and water.
Her mother took in washing.

'You're not very fond of my skinny friend?
I wouldn't want you to be!'

Gretchen shuddered. Her hand moved
To the crucifix round her neck.

'So you're wearing a cross,
You little darling!
I gave it to you?
Yes of course I did.'

He remembered Mephisto grinning
As he handed it over.
'When in Rome . . . ' the fellow had said.

'It's too good for you?
Nonsense, my sweet – nothing's too good . . .
Oh dear, the chain's broken –
Where can the cross have got to?
It can't have gone too far . . . '

It went too far.

Faust is forbidden to marry

'Remember, there's no marrying
For Bachelors of the Black Arts!'

'But how shall I put it to them?
The subject is bound to come up.'

'You could tell them that you're waiting
For divorce to be invented.'

'According to the book, it is
Better to marry than to burn.'

'The book doesn't say anything
About marrying *and* burning.'

Faust is soon bored

'What drew me so potently
Was her utter innocence.
Now it's gone, gone utterly,
With nothing to replace it.'

He'd eat his crumpet and have it,
Mephisto chuckled. Such a fuss . . .
'But what about experience –
Isn't that a pleasant bonus?'

'That takes time and talent,'
Griped the great lover.
'I don't have the one,
She doesn't have the other.'

                        *

Mephisto sought to cheer his friend
By telling how in years to come
In a far-off eastern kingdom
Subtle engineers would vend
An artificial maidenhead.

'Made of – I don't know what,
But sold as MATRIMONIAL-BOON,
INNOCENCE-REGAINED, SURPRISE-
YOUR-HUSBAND-WITH-A-SECOND-HONEYMOON . . .
One old fellow will be so surprised
His heart stops on the spot.'

By which date they'd be turning out
Artificial hearts as well, no doubt.

A game of consequences

It had rained the whole day long,
    And Faust was moping.

Mephisto said to Faust:
    'Let's have an evening out.'

Faust said to Mephisto:
    'As long as you're paying.'

So they went to Spicy Spiess's
    In downtown Wittenberg.

There Faust met a go-go dancer
    Called Meretrix, big bust, long legs,

Lusted after by university wits
    And city aldermen alike.

Faust said:
    'I thirst.'

Mephisto told him:
   'Try this Venusberg Lager.'

Meretrix said: 'I'm crazy about theologians!
   Make me an offer I can't refuse.'

Mephisto did, the handsome devil.
   Consequently, there were consequences.

Mephistopheles remembers the face of God

'When mankind fell,
They fell into unremembering.

Not so the angels.
When we fell,
We fell into unforgetting,

The flames, the ice,
Its stark unwinking light.

Do you suppose that I
Who used to look upon the naked face of God
Would not do anything
To look again?

Should a ladder stretch to heaven,
Its rungs all razor blades
That slice me to the loins
At every step –
Yet I would do my best
To climb that endless edge,

To look once more,
Then backwards fall
To timelessness in hell.'

'Don't make me cry!' scoffed Faust.
'You hope to curry favour up above?'

'I hope for nothing up above.
Time has rinsed my eyes.
I merely speak the naked truth.'
He knew he could afford to.

Faust already had forgotten.
This talk of naked made him think of Meretrix' bare thighs.

Faust suffers a temporary indisposition

'The girl's well named,' Mephisto leered.
'First merry tricks, then sorry pricks . . .'

'It's not a joke! I feel unclean.'
He thought of what his mother would say.

Swallows a camel and strains at a gnat,
Reflected Mephistopheles.

No rose without a thorn after all,
It was the price of admission.

'A pity you're not a musician –
They say it's marvellous for creativity . . .'

'You brought it into the world, you devil!
I command you to take it away.'

'Cocksblood!' swore Mephisto. 'Be a man!
I'll soon rid you of your little trouble . . .

    Shin Aleph Cheth Priapus
    summum bonum medicinae sanitas
      spirochaeta pallida
      inferni ardentis monarcha
    solamen miseris
        socios habuisse doloris
    pox vobiscum
      penicillum –

There, you're as good as new again!'

In the tavern

Oi recken tis a hurenhaus, no good Christchin as all they skivvies ter
do is biddin – Baint is biddin they does, tis is beddin – Oi eard tell
thikky rum kerl, im that looks loike a yard o pumpen-wasser, do be a
flittermaus or vampir – From oly roman vampir eh? ha ha – Meister
sends that dirne Gretchen ter fetch the brötchen, bäcker's boy won't
go there no more, not after a uge black pudel made a uge black
puddle on im – Bäcker's boy be off is loaf anyroads – Ar but there be
igh jinx up at doktor's ouse, they say as ow e be makin little babbies
in testtuben in is keller – Then what be all they jung mädel fowr?
twont niver catch on, could do yerself a nasty schaden wi that there
glas – Oive eard orrid noises at noight, speaking in unbekannte
zungen loike in church, Oi allus says a vaterunser when Oi goes by –
If yer can remember the words, yer betrunken ole bugger – Beim
himmel, ere comes is lordschaft isself wi that fancy furriner – Oi be
off ome, e played a schmutzig trick last toime, divel take im, turned
me wein inter wasser – Haw haw, yer can do that wiout no elp from
Ole Nik, yer silly pisswinkel –

Faust considers the subject of souls

Gretchen has a large soul and small breasts
Meretrix, great big breasts and a tiny soul
Breasts are of this world: Faust loves them
Faust loves not souls, especially large ones.

The Doctor's parents journey a great distance
to see their son

'Dear me, your aged parents have arrived!'
Mephisto winced. 'With lots of luggage too.'

Faustus embraced his mother gingerly.
He took his father's calloused hand with caution.

'I'll get the spare room ready,' Gretchen offered.
Said hearty Merry, 'We can all bunk down together.'

What an outlandish crew! His mother shuddered.
'The ever-womanly must raise or sink you, son.'

229

'At this time of the year it isn't natural!'
His father poked a peach, his finger went right through.

'We came because we've heard strange stories, son,'
The father said. 'Not nice at all,' the mother added.

'Gossip is a dreadful thing,' Mephisto sighed.
'Smallmindedness and envy,' Faust said airily.

'I don't imagine you'll be staying long?'
Mephisto didn't mean to run a lodging house.

Faust's parents question Mephistopheles

'Where's Miss Gretchen?'
'Gone to church, to pray.'

'Where's Fräulein Merry?'
'Gone to bed, to play.'

'Where's Master Faustus?'
'I wouldn't like to say.'

Mephistopheles addresses the working class. Faust's aged parents are nonplussed

'And your children's children
Shall learn how to spell correctly
How to pen a neat and legible hand
And to read good books

They shall be granted scholarships
And shall suffer therefrom
But they shall succeed in their time
Where their forefathers might not try

And they shall then discover
That correct spelling is held cheap
That the best people hire scriveners
And books are no longer read

Doffing their smocks and overalls
Your children's children
Shall go out to meet the world
And find themselves attired in last year's fashions
(For all things change and remain much the same)
And the world shall ignore them
As hitherto.'

*

'I can't see what all this
Has to do with our boy Faust,'
Said the aged father.

'I can't see what it has to do
With anything,' mumbled the aged mother.

'It hasn't,' grinned Mephistopheles.
'Your boy Faust has much more going for him –
He has me.'

'I wouldn't mind him being rich,
As long as he was honest,' said the father.

'I wouldn't mind him being poor,
As long as he was happy,' said the mother.

'Youth and age will ne'er agree!'
Sang Mephistopheles, gyrating
On his antique toe.

Faust betrays signs of forwardsliding

'Do you hope to emigrate internally?'
    asked Mephistopheles coldly.
'We have ways of tearing you to pieces.

What's done can't be undone.
What's said can't be unsaid.
Any jackass can repent,
But expiation – that's a literary conceit.
People write it,
They don't live it.

So make the most of what you have,
As we shall.

You cannot take it with you,
As we can.'

Mephistopheles calls upon his Master to preserve Faust against the influence of his aged parents

'Look upon this thy servant Faustus, who is grievously vexed with the wiles of a clean spirit, and grant him pardon of all his residual decencies and inadvertent kindnesses . . . Let thy servant be defended by the sign on his * of thy Name . . . Drive far hence, O Drol, the pious tempter, and send thy fear upon the tame beast which devoureth the Spillers.

Go out, thou friend of virtue, thou most wicked lamb, the author of marriage, the milkman of human kindness, I adjure thee that thou depart into everlasting temperate conditions . . . '

*

'I never thought to hear such language!'
    Said the aged father.

'It's what they learn at college,'
    Said the aged mother.

'I'm glad I'm just a humble husbandman,'
    Said the father.

'I'm glad I'm just a humble wifewoman,'
    Said the mother.

'We see that we're not wanted here.'
    The aged father fetched his knapsack.

'Wrap up well, keep regular, dear.'
    The aged mother sniffed a tear back.

Mephistopheles salves Faust's conscience

'A delightful old couple,
But it's for everybody's good . . .
Remember what the poet's bound to say –

> Oh how hideous it is
> To see three generations of one house gathered together!
> It is like an old tree with shoots,
> And with some branches rotted and falling . . .

We wouldn't want our house to be hideous, would we?
We wouldn't want it to rot and fall down?

In ancient Greece – well known of course
As the cradle of western civilization –
Old people were settled in the Underworld.
They could prattle to their hearts' content,
Drink gallons of tea, knit woollies, carve pipe-racks –
And occasionally relatives would visit them . . .

No call to feel guilty, cher Maître!
What was good enough for ancient Greece
Is surely good enough for the Middle Ages.
And that reminds me . . . '

II

Faust desires to have as his paramour that beautifull and delightfull peece Helen of Troy

'Sweet Helen, make me immortal with a kiss!'
Mephisto nods: there was no objection to that.

Golden tresses, swanlike neck and coal-black eyes –
But was she bold and wanton as bespoken?

'Thy lips suck forth my soul: see where it flies!'
Mephisto shakes his head: it should stay where it was.

'That's enough of this youthful idealistic stuff,'
He hisses to Helen: 'You know your business.'

She glares coldly at him, not yet finding her tongue.
She thought her function was to classicize.

Mephistopheles and the primrose

Mephisto smiled at the primrose,
The primrose smiled back.

'A primrose by a river's brim,'
Said Mephisto sanctimoniously,
'A yellow primrose is to him –
Faust, that is –
And it is nothing more.'

The primrose looked unsure.
'But what am I to you, sir?
I feel I can talk to you.'

'To me,' he replied judiciously,
'A primrose by a river's brim
Is something which to him –
I refer to Faust –
Is nothing more than a yellow primrose.'

The primrose dropped a tear,
Or perhaps a drop of dew.
'I just can't win, I fear . . .'

'Cheer up,' said Mephisto.
'I am sure there is someone somewhere
Who regards you as something more
Than a yellow primrose.'

'Oh?' The primrose perked up.
'And who might that be?'

'The Lord God,' said Mephisto.
'After all, he made thee.'

Adding thoughtfully, 'Mind you,
He also made a host of golden daffodils,
To mention but a few.'

The Doctor cannot abide gnats

'Phew!' Faust whistled.
'She's all woman, is Helen!'

'That's nice,' said Mephistopheles.
He was glad there'd been no muddle.

'But she'll have to put away
That Greek tunic.
It's far too revealing.'

'Really?' said his friend,
Who hadn't noticed a thing.

'I can always swallow a camel,'
Faust confided, man to man.
'A camel's a meal –
But a gnat in the throat is hateful.'

Faust and the daffodils

'Top of the morning to you!'
Cried jolly Faustus.

The daffodils tossed their heads,
They too were jocund.

'What a crowd of golden daffodils!'
He exclaimed. 'Or even a host . . . '

The daffodils tossed their heads again.

'One might say you were fluttering
And dancing in the breeze . . . '

The daffodils nodded vigorously.

'But mind your heads don't drop off,'
Said Faustus, bending down.
'Only the best will do for Helen.'

In extenuation of taking a concubine, Faust pleads vagueness

'An affair with an immaterial and possibly
Mythical figure hundreds of years old –
How could that get me into serious trouble?
If it were the Rektor's wife I could understand . . . '

'If it were safe there wouldn't be any fun in it,
Would there?' said Mephistopheles.
'Remove the sin from penis and there's not much left,
Is there? You might as well take Origen's tip.'

This is the fellow who calls me trivial,
He thought to himself.
'Do you really fancy the Rektor's wife?' he asked.

*

'Faust was right,' mused Mephistopheles.
'All this sex isn't getting me far,
And he's even enjoying it . . .

A shady lady with a shady past,
A go-go girl paid well and on the nail,
A silly miss called (save us!) Gretchen . . .

If only I could get him into bed
With one of his students!
But that sticks in his gullet . . . '

But then he brightened.
'Love of course is another matter.
Love calls for concentration –

A thing he hasn't time for.
At least, if I've not harmed him much,
I haven't done him good . . . '

Mephistopheles is taken with an excess of zeal

'Here's a reverend gentleman
Writing a history of witchcraft and demonology –

Demonology! – haven't we gone up in the world!

Now he's describing a "typical witches' sabbath" –
"One cannot write in dainty phrase of Satanists and the Sabbat . . .
This ostrich-like policy is moral cowardice" –

I don't remember any ostriches at our Sabbats . . .

The participants, he says, are subject to
"Mania blasphematoria and coprolalia" –
I wonder what that word means?
All in all it is a "Hellish Randyvous" –
Well said, dear Reverend!

That's what I call first-rate publicity,
Especially the medical bits . . .

I want it in every national and local paper!'

*

(Hundreds of years later it bore fruit
In bodies like the Witches' Liberation Movement,
The Witches' Anti-Defamation League,
And the Witches' International Craft Association.

But Mephistopheles took no joy in this.
Soon there would be no one to liberate,
Or even to defame. Soon there would be no infamy.
He declined to recognize the guilds,
He clung to the virtues of private enterprise.)

*Faust enjoys further instructive
conversations with his friend*

'So you want to do good?
Well, that is good of you . . .

The menu sounds bewitching,
The food tastes like a witch's brew –

Therein lies your difference
Of principle from practice.

To love an idea is easy,
Harder by far to bear with people.

Only as aesthetic phenomenon
Can existence be justified ? –

I don't know what you mean by that,
Unless it's dressing in pretty clothes.

But if your heart is set on doing good,
Don't let me stop you!'

*

'In time to come, you say,
The population will have swollen,
And evil with it.

Kind of you to worry for me.
I know I'm getting on,
An old limb of the devil . . .

True, I see an age ahead
When every card's a self-appointed ace,
And ripe for trumping.

But there is virtue in large numbers,
As well as vice . . .

No call for cataclysms –
A man can lose his soul in many ways,
Not all spectacular.

I have no taste for mass seduction,
Souls carted off on conveyor belts.

I rather see it as an art,
Damnation made by hand to measure.
I fear that I'm old-fashioned.'

Mephistopheles' ingenuity displayed

'Keep us informed of progress!' cried Mephistopheles,
Bursting into the back-room.
'Or even of regress – for we mustn't deceive ourselves.'
He was in the throes of a notion.

They should construct an electronic calculating machine
Capable of analysing the good and evil throughout creation
And providing as it were an up-to-the-minute scoreboard.

The demons were deep in a game of poker.
'Should we include a spiritual breakdown of the animal kingdom?'
Asked one of the players.

'Pray save your sarcasm for an appropriate occasion,'
Said Mephistopheles frostily.

'God sees all,' muttered another. 'Why don't you get together?'
But no one was very worried.
Mephistopheles' crackbrained notions rarely got past Lucifer.

*

During the Terror, a dedicated young anatomist was preparing
heads for dissection by boiling the skin off. A new batch arrived
straight from the guillotine. One of them was tonsured. It was the
head of a dear friend. (This was engineered by the witty Mephistopheles.) Secretly, while his colleagues were occupied elsewhere,
the young anatomist kissed its brow. It seemed to smile at him
tenderly. Then he dropped it into the cauldron along with the
others. He kept it for himself to dissect, and preserved the skull for
several years.

Progress or regress? – the demons wondered. Could that electronic
calculating machine have told them? They were often to wonder.

Faust is blessed with a son

Insubstantial though she was,
Helen found herself impregnate
With Faustus' son, named Justus.

'Justus bad as his dad,' Mephisto joked.
He offered to stand dogfather to the babe
But the mother turned her Grecian nose up.

'*Duty, obligation* and *spelling –* '
Announced the proud father, 'the very words
Shall be excluded from his vocabulary!'

Despite the attention lavished on him,
The boy turned out a rebel
With a perverse interest in orthography.

'This infant has no character,'
Faustus complained to Helen.
Mephisto wondered if he had a soul.

*

Later, he took a peasant girl to wife
And succoured his aged grandparents.
They all lived happily a long time after.

He was good with animals, and had green
Fingers, but neglected the Greek authors.
'I prefer true stories,' he would say.

Helen vanishes

'But she's fading away,'
Faust said in dismay.
'We must send for a doctor.'

Mephisto shuddered.
That was the trouble with shades,
They faded away.

'I can see right through her!'
Lamented Faust.
'Was this the face . . . ?'

Mephisto shrugged.
That was the trouble with cohabitation,
You soon saw through it.

Then one day she wasn't there.
'Gone to Paris for the weekend,'
Mephisto ventured.

But she never returned.

Walking in the Harz Mountains, Faust senses the presence of God

God was a brooding presence.
Brooding at present over new metres.
In which his creatures could approach him,
In which they could evade him,

– And he be relieved of their presence,
Through art as Proxy Divine –
Sublimation, as they termed it,
Which could very nearly be sublime –
For which he was truly thankful.

But how active they were, the bad ones!
They brooded rarely.
They talked incessantly,
In poisoned prose from pointed tongues.
How gregarious they were!
They needed friends to wound.

But who had invented tongues?
(One had to be careful when one brooded.)
And even the better ones
(One had to remember)
Were only human . . .
He started to fashion a special measure
For the likes of Gretchen, a still, sad music.

Creation was never finished.

Faust has a visitor

There was a rapping at the door.
'I expect it's for the Professor,' said Wagner.
'Could you take it, Herr Mephistopheles?'
Opening doors made Wagner nervous.

*

'. . . in a nutshell, my respected husband the hochgelehrte Herr
Rektor is a renaissance man as they say, his brow is very high and
his nose is always in a book, my poor feet are always on the ground
of course, when I talk of nature he replies with Nietzsche, I need
paraffin, what do I get, I get Paracelsus, he doesn't appreciate my
cooking, he reads at table, all this learning, it's bad for the juices,
he's a Latin lover, he reads in bed too, Virgil he says, more like vigil
I say, it's downright incompatible, he complains of crumbs in the
bed, what I complain of is his ologies and osophies, little use they are
to a woman in her prime and best nightdress – '

'This way, please,' said Mephistopheles. 'We have an opening for a
good plain cook.'

'How dare you, sir! Kindly remember that you are addressing the
gnädige Frau Doktor Rektor of Wittenberg, and as for plain, well let
me tell you . . . '

*

Eventually she came in.
'But I can't stay long, a woman's work is never done.'

Mephistopheles encounters the Blessed Boys

While minding his own business one evening
Mephisto was beset by a chorus of good little spirits.
They hopped and skipped around him, chanting.

> 'If the assault be keen
> Fearless must be our mien'
– And so forth.

'Showing their knickers!' Mephisto thought.
'What a terrible thing innocence is . . . '

Soon they were calling him names –
Senile, envious, frustrated, spiteful, impotent . . .
'Reactionary negativism!' squeaked one of them.
'Counter-productive!' piped another.

Obviously from a recent intake.
He gave them a horrid leer:

'You'll look silly when you're my age –
Joints swollen with gout, your haloes moulting,
Lungs gone to pot, and drawers drooping . . .'

He shouldn't demean himself by answering back.
He put out his tongue at them.

'Time for supper!' one of them twittered,
And off they scampered.

In the Senior Common Room

Poor old Schleppfuss, he'd just announced his subject, 'Beauty in a
woman is like the gold ring in a sow's nose', when in walked his wife
– By the way, has anyone seen Faust recently? – Is he not *corpus
naturale* and is not that *mobile*? – Oh, so he's away on sabbatical? –
Peripatetic is the word for Faust – Nothing pathetic about him, *wie
man sich bettet, so liegt man* – Something's eating him, bills falling due
I dare say – Which reminds me, when's the next review of salaries? –
The Rektor's in such a foul mood these days one doesn't like to bring
it up – Lucky for friend Faust he has tenure – Lucky that nobody's
ever defined that clause about *gross insobriety and impropriety* – Ah
well, *de minimis non curat lex* – How's your new book progressing? –
There's more money in astrology, it's the coming thing – What's the
weather forecast? I'm examining at Göttingen next week – Brr, it's
been so damned cold we've had to burn our laurel boughs – Yes,
time for another bottle of port, I think –

Mephistopheles is sorely embarrassed

'Please,' begged little Justus,
'Tell me what the soul is, Uncle.'

'A fish, my lad, fine and tasty –
You ate one yesterday.'

'You're playing with me, Uncle,
You treat me like a baby.'

'Right-ho, it's what you tread on
As you walk through life.'

'It must mean more than that!
I like true stories best.'

'So, it denotes a state
Of utter loneliness'
(His voice grew bleak)
'Where everything is lost
Save what you never wanted.'

'Forgive me, Uncle,
I cannot think that you are wholly right.'

'I'm not far wrong,' Mephisto snapped.
'Why can't you be like other boys
And play at tearing wings off flies?'

'Uncle dear, do flies need wings
Because they have no souls?'

Mephisto sweated. This was how it felt
To be a family man. 'Look, laddie,
This uncle is a Harmful Influence –
You'd better cultivate your Uncle Wagner.'

He turned and hurried off.

Mephistopheles relates a tale of olden times

'So that very night,' said Mephistopheles in a
    confidential whisper,
'I slipped into the bedchamber of Pilate's wife –
A remarkably handsome woman, I may say – '

'Yes?' said Faust, hanging on every word.
'And what did you do? Did you do it?'

'Oh, yes, I certainly did.
I snuggled up close to her, on the pillow –
What an opulent figure she had! – her name was Claudia,
    by the way – '

'Do get on with it!' Faustus was twitching.

'And I whispered a dream in her ear –
Exquisitely shaped it was, her ear I mean, like a shell,
    only much softer of course – '

'Never mind her ear – a dream did you say?'

'A trick of the trade. We tried it on Eve –
She had a fine ear too, sort of pink and innocent –
But we were caught then, we were only beginners.'

'Never mind Eve – what about this dream?
Didn't she wake up?'

'I put it to her – Claudia, that is – in fictional form
And with lots of strong emotions like you get in dreams,
That she should prevail upon her lord and master –
He was snoring in the next room, I was glad I wasn't whispering
    in *his* ear – '

'I do wish you'd hurry up with this story!'

'– and get him to stay the execution of Jesus and
Remit the sentence to thirty years of digging wells
    in the desert . . . '

'That was high-minded of you,' said Faust, abashed.
'I trust you got something out of it?'

245

'The idea was to avert indiscriminate redemption,
Of course.' His voice fell. 'But it didn't come off.
It seems Pilate's wife was always having dreams –
Fancy that ! he'd say at breakfast, but he never listened.'

Faust is troubled by a dream

Faustus dreamt a strange dream, strange and deathly.
In his dream he heard the voice of a small lady,
Piping perhaps, or perhaps chanting chillingly.

'Aroint thee, false witch!' cried she with a brave face.
'Human inventions help properly, magic is a disgrace.'

Still shaking, Faustus told his dream to Mephistopheles.
'It was a poem,' said the latter, 'and only meant to tease.'

In actual fact, he continued, the lady in question
Was the proud owner of a mighty and intricate engine,
A Washing Machine, which cleansed and rinsed one's linen.

'By no means pure magic, but merely applied.
Nothing at all to be scared of, my son –
Before long no home will be complete without one.'

'Human inventions are nicer,' Faustus opined.
'For Gretchen's a dab at wash-tub and flat-iron,
The Frau Rektor can cook a tiptop hotchpotch,
And even Meretrix is good at shelling peas . . .'

'But you are privileged,' said Mephistopheles
Indulgently. 'You have the magic touch.'

Faust appears before the Senate

Addressing the Senate of Erfurt,
Faust offers to restore various lost masterpieces
For as long as it takes to make copies.

Among them,
The missing comedies of Terence and Plautus,
The complete works of Sappho –

'I have four teen-age daughters,'
One of the senators starts to protest –

Aristotle's *Customs of the Barbarians*,
Homer's *No Second Troy*, or Post-war Developments,
And Cinna's verses for which he was slain.

'Very interesting,' says the chairman, 'I'm sure.
But our clerks are under considerable strain
What with minutes, motions and amendments . . .
However, we shall make a note of your kind offer.'

'Hear, hear,' mumble the senators.

Wagner is anxious

'What is amiss with my master?
Once the lecture theatres rang with his *sic probo!*

He rarely takes his classes these days.
He's always on the wing –
Off to an international conference
Or advising foreign governments –
Or closeted with that – that Levantine,
    I suppose . . .

Or else relaxing with a lady.
Well, that's all right, it's normal,
    I suppose . . .

But how I used to relish his *sic probo!*
I like things being proved.
I like plain honest truths.
One day perchance I'll marry a plain wife,
A nice shy ordinary girl.
Trouble is, I'm shy too – how shall we get started?

Sic probo! I shall say to her, and kiss her.
But firstly I must get my teacher's licence . . .
I wish my master hadn't grown so young.'

Gretchen at the sewing-machine

'My stomach is sore,
My peace is fled,
I wish I'd been careful,
I wish I was wed.

Once I would scold
A poor girl in disgrace,
Now it's for me
To hide my face.

I love him still,
My dear Johannes,
His clever words,
His polished manners.

It's not surprising,
When I love as I do,
That those more knowing
In love, love him too.

There's no one to help,
I can't tell my mother,
The priest would be shocked,
I'm scared of my brother.

He says his religion
Forbids him to marry,
If only I
Had the luck to miscarry!

Perhaps his crony,
That creepy fellow,
Could tell me of someone
Or somewhere to go.

My stomach is swelling,
My peace is fled,
I wish I'd been careful,
I wish I was dead.'

Faust exhibits his necromantic powers

At the Duke of Parma's court
Faust performed unheard-of feats of magic.

He summoned spirits to enact
Goliath slain by David,
Judith beheading Holofernes,
Samson bringing down the house,
The Vandals sacking Rome . . .

The courtiers were enthralled.
'Unheard of!' they exclaimed. 'Unseen as well!'

But Parma's Duke spoke sternly:
'This circus is for common folk!
Have you no power to raise our minds
With shows of nobler deeds,
But only rapine, death and ruin?'

Faust bowed his head in shame,
And packed his raree-show.

At the gate the major-domo waited.
'His Grace desires to see in private
Adam with his bride before the Fall,
Mars and Venus caught in Vulcan's net,
King David and his virgin side by side,
And Cleopatra's nights of love . . .
He'll make it worth your while.'

'Humble regrets,' said Faust, 'the show is over.'

Lucifer broods

'It was quite simple –
If God could be god
Why couldn't I?

All I asked for
Was equality and independence,
Primus inter pares –

With not too many *pares*.
A rotating chairmanship
Might have been the answer.

What happened to the question?

Better to reign in hell than serve in heav'n –
But better still to reign in both.

What, you might ask, is the point of it all?
Seduction, corruption, ruination –
All this hard labour I put in,
Day after night after day

– The solace of having companions . . .

Companions! Do these lost souls
Imagine they're as good as me?
Damn their eyes!

So I am the Spirit who always says No?
Once I was ready to say Oh yes!'

*

'I do wish he would stop brooding in my head . . .
Where are the aspirins?' Mephisto whispered.
'God knows!' yelled Meretrix.

The political achievements of Faust

A waste of time it would be, dear reader,
To itemize Faust's political feats,
For in effect they were no whit brighter
Than his infamously vulgar leg-pulls –
Like turning a horse to a bale of hay
Or invisibly tweaking the Pope's nose . . .
Granted, he assisted the Count of X
To repel the powerful Duke of Y
With empty suits of armour and false fire –
But neither Count nor Duke was gratified:
Honest war it wasn't, for no one died.

He furnished the Z Republic with a
Computer called Rational Government.
Its rule was just, taxes were abolished,
Crime dwindled, the economy flourished –
And the good citizens were bored to tears.
Vast Indian stretches he saved from floods,
Only to see them laid waste by a drought.
When he clad all the students in silk and
Velvet, they quickly returned to their rags . . .
Thus Faust did good, as he had wanted, and
Little good came of it. Angry he grew
And embittered, but Mephistopheles,
His faithful helper throughout, merely smiled.

Mephistopheles lectures in the Professor's stead

'Your attention please, Studenten and Studentinnen –
We now come to the final scene of this curiously set text.
I shall ask you to evaluate in passing the random contingency of
    metonymic association and the substitutive totalization by
    metaphoric reversal . . .
No, I shall not repeat that, if you can't grasp it first time
    you shouldn't be here.'

To his horror he spotted the gang of blessed boys
    fidgeting in the back row.

'Mountain Gorges, Forest, Cliffs, Wilderness –
All the machinery of advanced Sturm und Drang, you will
    observe, in a semiological mishmash of arcadian and religiose.
Holy anchorites are paradoxically said to be on the move,
    while a Pater Ecstaticus hovers in the murky air . . .
And what, pray, is a Pater Ecstaticus?'

'A happy father,' said one of the cursed boys,
    'who has just had his eldest son and heir.'

Mephistopheles continued briskly:
'Various grades of angels make their appearance, some young,
    some old, some less perfect, some more so, but all talking big
    as per usual . . . '

'Hey look!' shouted one of the dreadful youths,
    'We come next – let's see what you make of us!'

'The *seligen Knaben* or Boys' Brigade seem to be doing
    something obscure but nasty to a dumb chrysalis – a deplorable
    concession to public taste.'

'Old phoney! Doesn't know what he's talking about!'
Somebody let off a fire-cracker.

'Explain please to us the Magna Peccatrix,'
    asked a pale-faced student from Fribourg.
'Is it possible she is a referent purely symbolic?'

'Il ne s'agit pas d'une nouvelle affaire Marguerite –
That's certain. Look it up for yourself: Luke, vii.
The plain fact is – '
He paused to consider what the plain fact was.

' – that, despite a flimsy story line, the author has introduced a
    surfeit of characters and failed to distinguish between them – '

He took a deep breath.
'Mulier Samaritana, Maria Aegyptiaca, Una Poenitentium
    (what a mouthful!), the epicene Doctor Marianus, and a
    Mater Gloriosa, presumptive wife to Pater Ecstaticus.'

He sighed.
'You will have noted a deficiency of realism in the work,
    and a lack of relevance to the here and now.
And the scene now comes to an end, not before time, with
    an apt allusion to women and the ineffable . . .
This lecture also comes to an end, because I have a symbolic
    headache.'

The blessed boys were making disgruntled noises.
'Resign!' they were heard to advise, 'Resign!'

In sorrow Wagner takes his leave of the Doctor

Wagner was offered work at Heidelberg,
Coaching some student princeling.

Loath though he was to leave his master,
He had to think about his future.

'Che sarà,' he told himself, 'sarà.'
But tears were standing in his eyes.

'I no longer feel at home here, Doctor.
I was ever ill at ease with ladies,

And your strange friend gives me the shivers.
Last night as we played snakes and ladders

The board began to writhe and hiss –
"My favourite game," he cackled, "snakes and adders."

Forgive me, sir, but I must go from here,
And if I may, I wish that you would too.'

'So be it,' Faustus sighed and shook his hand.
'Then may you teach as well as you have learned.'

The prophecies of Mephistopheles

Of the worm Tyranny, which shall be chopped down
    and into a thousand pieces

Concerning the miserabilism of the Jews
    who may never turn away a pregnant female

That as Leben by Nebel, so shall Live be contained
    by its contrary,
And there shall be no life without file

Of the abolition of inferiority
    in so far as is consistent with superiority

Of the discovery of humanism,
    leading to a love of cats, dogs and aquarium fishes

That many shall drown in the depths of their skin

Concerning nations which speak with one voice
and likewise scrawl on walls

That when there is no longer room for lying down
mankind shall couple standing up

How alchemy shall yield to chemists, aspiration to
aspirin, barbs to barbiturates, aesthetics to
anaesthetics, whores to hormones, heroines to heroin

That even as men have created women,
so shall women destroy men

That a God all-merciful is a God unjust,
and that God will be just

How Faustus shall live for ever
and Mephistopheles a little longer

That only those prophecies are genuine
which show a profit.

Faust is interrogated by the authorities

A noise like thunder was heard at the door.
'Open at once in the name of the law!'

'It's open already,' replied the Professor.
'You need only apply the gentlest of pressure.'

'I am the Polizei Kapitän Scheister,
And with me is our esteemed Bürgermeister.'

'So pleased to meet you! Let's all have a drink.'
Mephisto snapped his fingers. 'Schnapps, I think.'

'We've had lots of complaints,' said the grave Polizei. –
'It's the weather,' sighed Mephisto, 'first wet and then dry.'

'I must ask this Herr to show us his Ausweis!
We're on the qui vive for spies in disguise.'

A cunning devil is never caught short.
'Take your pick – I've a pass for every port.'

'The neighbours complain about accents unholy!' –
'We're translating the Good Book into Swahili.'

'The game's up, Doktor, you might as well tell us
What you are up to down in your cellars!'

'Atomic mushrooms,' his proud friend replied,
'Coming on nicely – I adore them when fried.'

'And where is the mädchen by the name of Gretchen?' –
'Buried of course in the basement, mein Kapitän.'

'You think we policemen are boobies, ha ha!
She's really gone out for a walk, nicht wahr?'

'Brilliant!' cried Faustus. 'You're hard to deceive!
Now please mend the door before taking your leave.'

Faust desires to meet Lucifer

'I would dearly meet your Master.
Perhaps he'll throw more light than you can
On life's dark matters.'

'Lucifer brings light, but darkness too.
If you suppose that I am – as you say – designing,
You ought to meet my Master!
Lord of ambiguity, Shade of many a meaning,
Spectre of the spectrum, the Giver and the Taker . . . '

He found it hard himself to comprehend his Master.

'I'm the practical party. Straight and homely,
        though not insignificant.
But I expect you'll meet him one day, dearly.'

Faust is set upon by Gretchen's brother

One dark night Faust and Mephistopheles come face to face with Gretchen's brother in a deserted alley. He is a soldier, on compassionate leave, and called (such is the fancifulness of the lower orders) Valentine. 'Here's trouble!' mutters Mephistopheles. 'He's heard about his sister.' Valentine pulls out his sword. 'You swine!' he cries, among other rancorous and trooperlike ejaculations. Mephistopheles tells Faust, 'I'll keep him busy, then you stab him, preferably in the back, with – what do you have on you? – a quill, a piece of chalk, a forefinger – anything will do the trick.'

Mephistopheles and Valentine begin to fence. Clash, clash! Mephistopheles makes a sign, and Valentine's sword-arm is instantly paralysed. 'Magic does have its uses,' Mephistopheles pants. Faust steps bravely forward and runs Valentine through with an outstretched finger. Valentine falls, crying out, 'My sister – a strumpet!' And dies. 'Officer,' says Mephistopheles in a posh voice as the Watch comes round the corner, 'there's a poor fellow lying here and wishing for a trumpet – to play the Last Post, I fear.'

*

'It was two against one,'
Moans the Doctor,
'Unfair competition.'

'All's fair in love and war,'
Says the other,
'And this was both. What's more,
He was a professional
While you're an amateur.

Better a living amateur
Than a dead professional.'

Faust finds it is later than he thought

'Not twenty-four,' said Mephistopheles,
'But twelve years only if you please.'

'The pact,' Faust shouted, 'calls for twenty-four,
As I can prove in any court of law!'

'Your calculations must have gone astray –
You used my services by night as well as day.'

'You're nothing but the cheapest sort of cheat!'
Faust cried.

'A cheat I well may be, but I'm not cheap,'
His friend replied.

'By night I slept,' Faust said, 'and so
Did you, you dog, for all I know!'

'Indeed you slept – you slept with Gretchen,
With Helen, with the go-go girl from town –

You always rose to the occasion,
Because I never let you down . . .

And if you only dreamt at night,
Who made your dreams but I? –

Oh, I worked double time all right!'

Faust asks himself what he has gained

'The fool, they say, persists in his folly.
So does the proud man in his way –
It may be folly, but it is his own.
While still man strives, still must he stray.

And if the proud man throws good money
After bad, at least this will proclaim
His disrespect for riches,
Or prove the justness of his aim.

To save his intellectual honour
The clever man cuts off his nose –
Who knows, his noseless face may grow in stature,
Or else some grander nose may grow?
While still man strays, still shall he stride.

257

I told this hound of hell,
This pitman paid to help me strive,
That I would learn
        why it is the guiltless suffer,
        and why the wicked thrive –

His answer was: "As you discern,
God moves in a mysterious manner –
Unlike me. Me you can understand."

But now I think that I have understood
Whether or not the guiltless always suffer,
Not all the wicked always prosper.

And I shall write this in my blood.'

Faust makes his will

'After the specific legacies herein entered
    have been disposed of,
I devise and bequeath the residue of my estate
To Madam Meretrix for her absolute use and benefit'
(Meretrix saw to that)

'My second-best bed I bequeath to Fräulein Gretchen,
    of address unknown'
(She lived far off, with a man who wanted children)

'To the Frau Rektor of Wittenberg University
    I leave the Herr Rektor of Wittenberg University'

'To Master Wagner, my faithful clerk,
Whose counsel I have cause to regret ignoring,
    I leave my books,
Among which are some he is entreated to burn,
Also my ink-horn and my gold hunter for which
    I have no further use'
(*Terminat hora diem*, Wagner would say, tapping it)

'My son Justus Faustus I cut off without a penny,
    for the good of his soul'

'To my incorruptible parents
   I leave their pleasanter memories of me'

'To Johann Gutenberg, if he can be found,'
(He could never be traced)
'I demise the sum of twenty-four guilders,
   for it must seem his guilt'

'A certain immaterial item has been assigned
By separate instrument to Herr Mephistopheles:'
(Much good might it do him)
'Being without prejudice to Madam Meretrix' interests,
   this need not be specified here . . . '

*Mephistopheles assures his companion
of immortality*

   'Why did I not give up the ghost
   When I came from my mother's belly?'
   Thus lamented Faust.

                  *

   'But you're lucky, Doctor,
   You are privileged,
   You'll be remembered –

   Not exactly in flowing cups,
   But in story-books,
   A play or two, an opera perhaps . . . '

   (He saw the stage machinery running
   Amok, dragons and heaven collapsing,
   Killing two or three of the actors –
   Yes, that was at Lincoln's Inn.

   And in a town called Exeter,
   One devil too many on stage
   Who didn't seem to be a player –
   He would enjoy that.)

'In fact a noble Engländer
The Duke of Ellington
Will put you in an operetta
With a cute mulatto Helen!

Pity about your name though,
Doesn't have much of a ring to it.
Mephistopheles now –
There's music for you . . .

But you're fortunate, Faust,
You should count your blessings –
Most lost souls don't rate an epitaph.'

Faust's farewell to his friends

'We'd all had formal invitations. Must be something weighty, the
Dekan remarked as we went in. He started badly, I'd never heard
him stutter like that. My trusty and well-beloved friends, he said at
last, Hear me with patience and take warning from my story.
Naturally people began to titter. He mumbled something concerning wicked exercises and conjurations, and how his mother was right
. . . and his contract, and murder and lechery, and being wedded to
Lucifer. The Rektor walked out at that point. Maybe it's poetry, my
neighbour whispered. But it didn't rhyme. There was an hourglass
in front of him. The sands are running out, he cried. By now most of
the audience had too. I heard him say distinctly, Hell has many
mansions, each with a single bed. Then he howled, He will fetch
me! and fell against the lectern, crashing to the ground with it. I
hurried to him, but that weird friend of his was there first. He
growled something about drink on an empty stomach, hoisted Faust
over his shoulder and rushed off with him. Yes, it was a sorry
business.'

Lucifer advises Mephistopheles of the device known as sprats and mackerels

'Someone in Weimar has tacked together
An artful defence of your friend, I gather.'

It was Lucifer speaking.
He was speaking out of a cloud.

'It seems that Faustus was only striving –
And to his Lordship striving is pleasing.'

'I strove – and he succeeded!
That tale won't cut much ice in hell!'

'I think we yield to the cunning Geheimrat,
And suffer your fortunate friend to depart.'

'You want us to look foolish, Lucifer?
How can you do this to your servant? –
It's devilish!'

The cloud that Lucifer spoke out of
Grew darker. So did his voice.

'We all have our cross to carry,
But crosses are ladders.
We let the Doctor go scot-free,
To encourage the others . . .

Experto credite!'
The storm-cloud drifted away.

Mephistopheles wonders

His Master had too much on his plate,
Mephisto reckoned. It was beyond a joke –
Couldn't see the trees for the forest fire,
Sometimes spoke in a cloud of smoke.

Yet did Mephisto want Faust?
Maybe Rachel was worth those years
    of Jacob's labouring –
But the soul that could be purchased
Was rarely worth the having.

'If it had been Justus, now . . .
Or the aged parents, he or she,'
He shuddered slightly.
'Or even Gretchen, even Wagner . . .'

It was always the one that got away
You ended by wanting.
Sometimes he thought his work
Wasn't very satisfying.

Sometimes he wondered
Who he was working for.

The commander dismisses his troops

The times were out of joint,
Ancient custom gone by the board . . .
The troops were standing to.

'Stand down, platoon!
Fat fiends with blunt horns
And lean ones with broken ones . . .
No, stand up, you lumps!

The mission has been aborted.
It's roses roses all the way for Faust . . .
So hand in your pitch and brimstone –
And wipe those looks off your faces!

No muttering in the ranks!
Yours not to reason why.
If action's what you want,
Then go chase the blessed boys –

Or girls for all I know, the Gestümper!
And find out what they're made of,

But remember, just rough them up,
We don't want a diplomatic incident.'

Poor devils, he thought.
It was a question of morale.

Faust and Mephistopheles make their adieus

'We too can move in a mysterious way,'
He ground his teeth. 'Once in a while –
So don't pin your faith on it!

I shall tear this contract into pieces.'
He did so, very slowly.
'You're free. If free's the word . . .

I scarce expect a testimonial,
But can't you make a special effort
And just for once say Thank you?'

Faust thought of how his hand had hurt.
'Thank you,' he said at last.
'I'm not entirely taken by surprise.'

'Indeed?' The other strove to look
Entirely unsurprised. 'Mind you,
You only gain what we are glad to lose –

Time, and time hangs heavy.
You'll miss me, Faust. I was your muse.
Who will amuse you now?'

The one considered things to do in time.
The other thought of things that time could do.
The stars stood still, the clock marked time.

'That's true, my friend,' Faust sighed.
'You've ruined me for other company.
I'll pass my days in some far wilderness.'

Was there no wilderness at hand?
'I'll miss you too,' said Mephistopheles.
'But then, I'm used to missing . . .'

POEMS 1981

Sketchbook

Having made a tree, one creates another
For the sake of something called symmetry.

Between the trees
There had better be something to see –

The sea, the sea! But what are trees
Doing in water? That's how the rot sets in.

So between the trees one plants –
What shall we call it? – dirt . . . soil . . . earth!

And on the earth
There had better be something to be –

A being! Then on behalf of symmetry
One makes another. That's how trouble begins.

Lingering Parents

When you could do whatever you wished,
    except one thing alone,
What was the thing you wanted to do?
(By our fruits are you known.)

The crops just grew.
The animals were backward, but loving.
When one of them spoke, you naturally found
    its discourse improving.

Yours wasn't the first lie. Only the second
    and so on.
The snake had ensured a distinguished future,
Now you had to think of your own.

Less loving the beasts now, less backward too;
You covered yourselves in clothes to remove,
    for so were both of you,
And the weather changed for the worse.

Winged Shapes were your introduction
    to swords being brandished,
To dispossession and deportation,
And dreadful faces.

What a lot occurred on that final day! –
Though the woman had time for dreaming.
It wiped out the memory, almost,
    of that unconceived beginning.

And you discovered how to suffer –
A field in which only the serpent
    had once forerun you,
By a length which nothing could measure.

Tongues, like the apple, were loosened,
    and someone said later,
'Who knows how to suffer is heaven's heir.'
There were worlds elsewhere.

Midstream

Half-way across the racing river
The big man groans: 'So heavy though so small –
You bring my life in danger.'

'Here is no wonder,' says the child
Behind his shoulder. 'See, you bear the world
And all its sins as well.'

The other halts. 'What credit do I gain
By this? How long my fame?
Wiser to drop you where I stand.'

'Go on you must,' the child replies.
One thing at least he knows for certain.
'I was not born to drown.'

The other sighs: the world shall have its ride
Then, here's no place to bandy words.
He bends his strength against the flood.

Development

The house of God is due to be converted.
These days He has no need of so much space,
The children all grown up and moved away.
A family scattered, sad but commonplace.

Bowling or housey-housey? No theatre
Or other devil's playground, heaven forfend!
A last stroll round the old backyard, and then
He leaves to room with some old Reverend.

Pains

There is something that aches from time to time,
Which might merely be lung cancer or muscular rheumatism.

One of the peculiarities of the word 'soul',
Musil remarks, is that young people cannot pronounce it
    without laughing,
And the middle-aged fight shy of it except in phrases like
'X has a noble soul' or, more commonly, 'Y has a base soul'.
It is, as he says, 'distinctly a word for the elderly' –
Many of whom prefer to talk to themselves.

We must leave Musil aside, for he was not invariably
Nor indisputably serious. He inclined to a chronic irony
Which led to the downfall of the Austro-Hungarian Empire.

How stands the empire of the soul? Declined or even fallen?
At best it is something one possesses oneself, but (seeing
One doesn't wholly believe in it) others most likely do not.
For the benefit of the doubt is given more readily to oneself
Than to others. One is more deserving, perhaps. Or nearer.
And, whether of growing or dying, the pains are one's own.

266

But this is the kind of irony responsible for the ruin
Of the Dual Monarchy. It is likewise enervating to wonder
Whether by absent-mindedness or insufficient conviction
The corner was created but not its occupant –

To suspect that the pain, located in what is traditionally
(There was sense in those old traditions) the proper place,
And feels like a longing, like a thing that wants to get out
And go elsewhere, which sometimes seems to be calling to
Similar bruises in other breasts, as if there was fellowship
Somewhere, and the best does exist, or did, or could later –

Enervating then to suspect (though not without reasons)
That it is only, by one of life's cheap little ironies,
Lung cancer or, by a still cheaper one, muscular rheumatism.

If the soul is uncertain, more slippery still is belief,
Which we used to think simple, the stuff of a trenchant sentence
    or two.

Those First Moments

How far does one's care for the body carry?
Somewhere in some way a line has been drawn,
Short of the flames, short of the ossuary –

But what of those first moments, when the soul,
As often envisaged, flutters uncertainly,
Ashamed to twitch at a tether unfastened?
The body is not discarded so easily.

In the East, girls learn to compose their limbs in bed.
One's body, lying in the street, would like some decency.

As the young were told: don't put on pretty undies
For an outing, you may be tempted to remove them –
So the old reflect: a perfect stranger can undress
You any day, cleanliness is next to . . . something.

The body slumps against low railings.
Here's a fine fast-flowing river, hums the soul,
Peering out of bleary eyes:
Water answers several questions . . . washed and willing.

The body straightens up, as if accosted.
Easy on, the soul says, can't you take a joke?
But neither laughs. In silence and in step
They make for home, a hot bath and scented soap.

Natural Species

There's a law these days against the extirpation of a
Natural species . . . So John Brown assures himself
As he moves with care down the Underground corridors.
A poster for panties carries a sticker:
   'This degrades women.'

For Brown himself is the sole survivor of one such
Natural species: the John Browns. He can still recall
The others – John the stripling . . . Brown the poet . . .
The sportsman . . . the bridegroom . . . the warrior . . .
   the work-force . . .

All departed, leaving this world a poorer place.
Only Old Man Brown is left, the last of his kind.
So will it preserve him, the law concerning species?
On a picture of men's briefs a sticker proclaims:
   'This degrades yellow underwear.'

He knows what laws are like. There are loopholes in them.
(They changed the rules of rugger when he wasn't looking.)
You can fall through a loophole. Or under a train.
Then what use a lawyer? He knows what lawyers are like.
On a playbill for *Oh! Calcutta!* a sticker complains:
   'This degrades Black Holes.'

Is his the folly of age? There are other forms.
Like the madness of Darwin. And the law's loopiness.
He moves with care through the Underground alleys –
Genocide they call it, or (watch out!) *Endangered.*
Better to miss your train than be hit by it.
   Extinction is degrading to John Brown.

Going on with
(M.C.G.)

'Of those things,' this book will tell,
'The which make death be arable . . .'

A place of cultivation, a fruitful condition,
The rich vale where the made souls ripen.

But some strange thing has happened,
It was the printer's hand that faltered.

For 'make death bearable' it now appears –
Is this the only fruit death bears?

Meanwhile we welcome little mercies,
A something to be going on with.

Burden

The grasshopper was a burden to me.
It knew of something hurtful to me.
In a dream I squashed the grasshopper.

Why was the grasshopper such a burden?
Its singing hindered me from sleeping,
All flesh is grass was still its burden.

Unlike the owl, the bat, the loris,
The grasshopper is no power of darkness.
It sings at ease in the light of the sun.

Did I lie at ease in the light of the sun?
The grasshopper hindered me from sleeping.
If so, then how could I be dreaming?
The grasshopper is no more a burden.

Guest

Is the kitchen tap still dripping?
You should always chain the door at nights.
Soon the roof will need repairing.
What's happening these days at the office?
Too much coffee agitates the nerves.
Now don't forget to spray the roses.
Do see the doctor about those twinges.

But tell me where you are! How is it there?
Are you in pain or bliss? And what is bliss?
Are you lonely? Do we live for ever?
How do you pass the time, if time there is?
Does God exist? Is God loving?
Why must his ways be so mysterious?
Is there a purpose in our living?

*

Why won't you speak of things that matter?
You used to be so wise, so serious.
Now all our talk is roofs and roses
Like neighbours chatting at the corner.

Here wisdom is as common as the air,
Great matters are the ground I tread.
Tell me, what weather are you having?
Are the planes still noisy overhead?
Ask my old mates how work is going –

Don't be angry, dear. This hasn't changed:
Those things we lack are what we covet.
I am the guest, the one to be indulged.

K. on the Moon

'If his parents love him so much, why
don't they give him 30,000 gulden so
that he could go off to some cheap
little place on the Riviera . . . '
      – Max Brod to Felice Bauer

Why is K. going to the Moon?
Because he is the thinnest person he knows
Because he neither smokes nor drinks
Because he does not care for communal life.

Off lifts the tin-beetle
Trailing its fiery slime
It looks good
(Except you can't see it).

But K. can see Fräulein Bauer
He can see Kakania, Krim and Korsika
Kreta, Kapkolonie and Kanada
They all look much as usual, only quieter.

'Dear Fräulein Bauer
I am an erratic letter writer – '

The typewriter floats above his head
Such things never happen in the office.

'Yet a letter is better
(Especially fine handwriting today, not so?)
If I arrived in person you would find me
Insufferable, and the thinnest person you knew.'

His burden he knows is weightless
But does his burden know?

'Will you marry me, dear Fräulein?
Will you marry a man who is sickly, taciturn, miserable and selfish
A man with white hair, poor prospects and palpitations
A man who is chained to invisible authorship
A man for whom even night is not night enough
A man who lives on the Moon?'

And would he want to marry a dear Fräulein –
Or anyone else who was willing to wed him?
Engagé he is but his health
Is not good enough for marriage.

'For a whole week I have seen nothing but
Outlandishly dressed Montenegrins in my sleep
Which gave me headaches . . .
When I recover my typewriter
I shall write a story called The Man Who Disappeared.'

He does not have a bent for writing
If you please, he is bent with writing.

'You read my letters carelessly. I said my uncle
Was in Madrid: you have moved him to Milan.
Is it all one to you whether I am on the Moon
Or on Mars . . .?'

The roads are long and it is easy to go astray.
He has acquired great experience in complaining.

'What have I done that you torment me so?
No letter today, no letter yesterday
And I believe it is now tomorrow.
Some diabolical official is playing about
With our letters . . . '

He runs in circles like a squirrel in a cage.
What might he not accomplish if he were free
And in good hands?
(Nothing.)

'Love me, Felice! And hate me!
Why do I torment you so?
You are my own self . . .
You are my human tribunal.
Please send a rose in forgiveness.'

*

A small step for mankind
But a giant leap for K. –
He falls on his face on the moon-rocks.
Have no fear, Felice –
Devised solely for writing, he cannot break.

Alas, the Montenegrins are there already
In their headachy garb!
A long thin butcher's knife glints in the moonlight.
Someone has been telling lies about him.
He is virtually a vegetarian.

We have a problem
(We always have a problem).

'Dearest F.
Now quickly forget the ghost that I was.
Peace will require no letters.
Try to drink less tea in future . . .
Kisses in the corner – K.'

The Moon is made of literature.

Missionary Unfrocked

We were the Lord's own locksmiths.
To deliver the lands of the heathen
From error's chain was our duty.
We could track down a fault most surely.
They hung us with garlands of jasmine
And sweat, and their women, held cheaply,
Who held us so dearly.
We grew less sure of our native skills.
We took their vices to add to our own,
Their virtues more rarely.

Now we dream of stale frangipani,
Of the monsters we hunted who caught us.
Our fingertips go hungry,
Unless some pale shadow accosts us:
'Are you feeling lonely, dearie?'
And we take her home to wind in a sari.
For we are brought low by our losses
Of grace, and scope, and heresy.
The Lord has taken his keys back,
We shall find no delivery.

Combat Camera

You could get anything there
You could get laid or opium or beer
You could catch your death

Once the dead were tallied in foreskins
Later in hands
(You could get laid, you could get hand-jobs)
Then in percentages
(Members of the body politic)

The wonder is the place has lasted
Has it lasted, is it there still?
You could get some stunning pictures.

Remembering Angkor

A smell of rotting vegetation –
Bats living in its cities,
Rope-veined trishaw drivers,
Prostitutes, those young antiquities,
Anything that crouches,
Beggar, soldier, dog or cripple –
All part and process,
Swarming, wasting, with the jungle.

Dead bodies will have fitted
Well among the ruins.
The half-starved children too,
It has to be admitted,
A few new ruins.

But some things couldn't be in tune:
A Pepsi-Cola stand, or equally
People's Limonade,
A fish and chip saloon,
A noodle stall or like amenity.
Some spots have their roots in savagery.

World in Action

The rioters are tearing down the prisons.
They think the Revolution has no need of them.

We however know precisely what will happen.
We know better, we have seen it all before.

Such knowing cramps our style. We look
To heaven and hell for something truly new.

Until which time we seek our small distractions –
In things that brought the Revolution into being,

So says the Revolution, pausing in its business.

Explanation

It develops like this, you see. The things called hands
Which terminate in fingers, which terminate in nails,
The whole depending from arms. And likewise the legs,
Which merge into feet, from which emerge what are known
As toes. There you see a head. These parts grow together
Quite slowly, or grow one from another. As though,
It might seem, a loving care is somehow involved.

A bomb, as they term it, is made of parts, quite slowly,
And with loving attention. It is then laid carefully
In places where arms and legs and heads are accustomed
To gather. A loud report ensues, and the parts
Remarked on above are disjoined, along with a reddish
And lately interior fluid referred to as blood.
What has come slowly together is swiftly taken apart.

Why is this? It is because of something not to be found
Among these pieces, an item unseen, which perhaps has
Escaped. It is something described as a soul, or a mind,
Or (by transference) a heart, in which are present things
Equally invisible, called thoughts or feelings or likes
Or dislikes. Where do these troublesome elements arise?
Their maker is invisible also, and very hard to locate.

Still Waiting

One doubts they are any sort of solution, those barbarians,
Whether they come dressed in long linen with short laws
Or worth their weight in gold and mystic abstractions,
Clad in denims embroidered with sharp and blunt instruments
Or as boy-generals, not knowing (how unspoilt!) their own ages.

(It is they, by the by, who bear the expensive walking-sticks.
One must trust they are intended only for walking.)

One doubts they have any solution to offer, these persons
(Few of us nowadays can speak either Latin or Greek,
But certain expressions increase in offensiveness),
Except the solution one finds in promiscuous burials –
Or, if you are fortunate, in a slight change of boredom.

It is amusing to lament, poetically, their non-appearance.
Unwise to assume that their likes will never arrive.

Animal Kingdoms

Catechism

It used to be held that birds had wings
Like angels and had taken no part
In the primal crime, and hence they lived
In the sky, near to God.
                                  Later it was shown
That they possessed two legs like men and like men
They fell.

Into our mouths? That was no fault of ours.
Besides, the mangers are full of dogs.

Dogma

Cats have four legs like us but
They do not speak in good plain short sharp terms
Like we do.
              It is cats that give dogs a bad name.

Their bite is worse than our bark.
They think they have nine lives –
But we shall have our day.

Bird's-eye View

Dogs are hogs. Cats are rats.
As for men – sighing and sobbing over our bodies
But still killing two of us with one stone.

Birds are at the top of the tree.
We have poets by the nestful.
Our brains are enormous and we are much prettier than girls.

What does little birdie say? –
Just wait till we have flown away!
You would do well to watch us.

The Question of Lust

When I was a child Lust was confusing.
One didn't quite know what it was.
It went on in higher circles. Like cocktails.
It cost money.

Then at school I began to learn German
And Lust was fine! It was
        Pleasure, Delight, Merriment.
The Jephson Gardens, packed with large peonies
And old ladies, was a veritable Lustgarten!
(But perhaps it went on in other circles?)

Lust was also a young lady, I found –
Struggling with Valéry with little success
In a Lusthaus in the Lustgarten of Leamington.
Lust led me to Faust
Whose secretary she was:
'J'ai tout ce qu'il faut, papier, crayons . . .'
It would be nice to have a secretary.
They go on in higher circles, and are often
Very comme il faut.

'You are not here to understand, mon enfant.
You are here to take down as I dictate.
And also to be not disagreeable to look at.
You understand?'
        'Since I am not here to understand . . .'

The Demoiselle Lust de Cristal
Was there to take down his Memoirs
Describing intellectual circles.
Research into time past or lost or even wasted,
It all costs money.

Now I am grown up and it is still confusing.
Lust and Lost, Faust and Proust, Past and Waste –
Most of it going on in higher circles.

Sex Shop

False noses, pimpled, with moustache attached . . .
A sleeping beauty waiting for the kiss of life . . .
Electric drills, perhaps, and cartridge belts –
Sophisticated weaponry for such small wars . . .
Nymph, in thy orifices be my sins remembered.

And back they flash, the sieges and the sallies
Of his early years. Hand to hand they were,
Those brisk engagements . . . Why, in those days
Even iron jelloids came in plain wrappers . . .
The brilliant memories catch his breath.

A young miss whispers, 'Look at grandad there!'
Her boy lifts leaden eyes. 'Must be his heart,'
He mumbles, feeling for his own.

Anecdote from William IV Street

Entering the publisher's warehouse, a foreign young lady
Asks for Volume XXIV of The Complete Works of Freud.

(This being the Index, at last, which directs the reader
To a wealth of unconscious wants he might else overlook.)

'I also desire,' says she, extending an elegant arm,
'An image of Jesus Christ approximately this high.'

I am a Bewohner of an Elfenbeinturm

It seems I live in an ivory tower
Or if I am German (am I a German?)
An Elfenbeinturm is my abode.
Do I abide in an elephant's leg?
Am I a flea then, a Faustian *Floh*?

If I were Chinese (but then I am
Not, China is not in Europe as yet)
I might reside in a Jade Pavilion.
Were I a more dynamic dreamer
I might inhabit a marble hall.

But if I am German only in part
(My German is only partly good)
Perhaps I dwell in an elf's left leg.
Or Number 11 Leg Towers maybe –
It has the ring of authenticity.

That I live at some remove from life,
Its down-to-earthiness, its low *Geschmack*,
Is owing to lifts forever out of service,
Trash on the stairs, the price of real ale,
And (oh!) the cost of public transport.

But write to me (if you can spare the postage)
At Flea Flat, Leg Towers, Elephant and Castle,
And tell me about the real and earnest.

Cargo

One day, your ladyship, it came in, my airship –

With barrels of strong drink and bales of tobacco,
Recordings of carefully selected vocal music
And a four-way stereo with two buttons only,

Home movies (mainly foreign), a Home Doctor,
An Order of Merit (for services to be rendered)
And a liberal gentleman's library,

An assortment of ladies and a lady's-maid,
Carboys of royal jelly and mountain oysters,
Also vulgarities like tea-bags and matches.

Then there fell from the sky a flimsy crate,
Long and narrow and on inspection empty,
A spade attached. The lady's-maid is a sturdy body.

But where were the gods, who should have dropped
On brilliant balloons, some time ago,
To bless, or to curse, or both, the cargo?

Was it worth it, wrenching open all those boxes?
Life is one long tea-room, with *The Desert Song* LP
For savoury. Only the lady's-maid is happy.

Ageing Sage

With the years their laps grow bonier –
My person happily grows plumper.

And even the arthritic can open tins
(The toothless can cope with the contents).

Moving around less, one is less trodden on
By those who move around less.

Something of a thinker, you venture?
Fear of the bow-wow is the beginning of wisdom.

Many have been the seats of my meditation,
Much is owed to support from public bodies.

Yes, I am pleased to recline on your lap.
But no, I have no intention of teaching.

Silence is the foundation of philosophy.
If you keep still, you are welcome to share it.

Orthography

The good souls spelt it 'abhominable',
As if to man's nature such things ran contrary.
A fallacious etymology –

Scholars laughed at that vagrant 'h',
Which the vulgar habitually
Omit wrongly or supply erroneously.

A thing 'of ill omen' and best avoided
(Though not to man's nature entirely contrary) –
It was 'abominable', properly.

Since nothing inhuman is alien to us,
We match the spelling with the derivation –
Without vain aspirate, vain aspiration.

The Retired Life of the Demons

This was the fifth heaven, the Angel told him –
Whereof Satanail was warden, from some height fallen,
A man of sorrows and expert in social problems.

'Metaphorically speaking,' spoke the Angel,
'These are but similes and metaphors.
They have laboured long and are worn out.
Literally speaking, they are greybeards.'

Stretched out huge in length,
The aged demons sprawled on rocking-chairs
Like companions in a fall.

Satanail raised his creaking bulk.
'Clichés, do you mean, that work damned hard,
Are never pensioned off, and last for ever . . . ?'
He shuddered vastly.

The Angel looked unhappy. A man of bliss,
He was subject still to embarrassment.

'We do not envy our nephews the glories in store
For them,' Satanail remarked. 'In fact, Mr Enoch,
Even autumnal leaves have something to contribute.'

'To be sure,' said Enoch. 'The fruits of experience . . . '
The Angel fidgeted. He was not at home here.

'Too old to stand, we sit on committees –
And of late, for instance, have devised a scheme
To temper homosexual activity in boys' schools
By introducing the heterosexual . . . '

The Angel turned his face away.
There had been angry talk of this, elsewhere.

'Conversely, and according to local needs,
To contain pregnancies in girls' schools –
The promotion of sodomy in schools for boys . . .
Small beer of course, but – how do you say? –
It all finds work for the devil's idle hands.'

'How interesting,' said Enoch on a note of interest.
The aged demons perused their horny palms.

Later they sat down to a light collation. Grace
Was gabbled by Satanail. 'Amen,' the Angel mumbled.
He had no such leisure to look forward to.

'Speaking of fruits . . . ' Satanail turned to Enoch.
The occasional visitor raised his spirits.

A *Priest in Balvano*

His house falls on His children.
He moves in a mysterious way,
As earthquakes do.

Then why were some preserved
From that far better home,
A heaven which never falls?

The reasoned ways of men,
All mystery removed –
Are human ways more welcome?

Are there sermons in these stones?
Or only flesh and blood and bone,
Once their maker's image.

He made us in His image,
Then gave a taste for reasons.
Then give us reasons.

Poetical Justice

It will be many years before I read again
Of the death of Cordelia,
Or indeed (though he deserves cuffing)
Of the Macduff boy's stabbing.

(Instead I shall read of the dissolution
Of science-fictional monsters,
Or the gonorrhoea of modern heroines.)

That such things happen in life is no cause
For them to happen in literature.
If it is true that 'all reasonable beings
Naturally love justice',
Then where shall they hope to find it?

I prefer to hear of such unlikely events
As Hermione surviving in private, or
Isabella furnished with a ducal husband.
(Some have tried, too late, to save Cordelia.)

In life we barely choose our words even,
Only those we hurt will still recall them.
Art, they say, continues life by other means –
How other are they?

INSTANT CHRONICLES

A Life

In biography, the fiction parts should be printed in red ink,
the fact parts in black ink.

Publishers Weekly

There is no life that can be recaptured wholly; as it was.
Which is to say that all biography is ultimately fiction.
What does that tell you about the nature of life, and does
one really want to know?

Bernard Malamud, Dubin's Lives

Lives of the poets

'They are present in the radiance of their works . . .
Like lights that blaze so bountifully on summer nights from
  bowers and lawns,
Like stars on earth, or diamonds, emeralds and rubies
Left in the bushes by an emperor's children at play,
Or raindrops hung in the high grass, sparkling with joy . . .

O never let us scan their lives at close quarters!
Forbear to seek those stars, those jewels, those raindrops in the
  cruel light of day –
All we shall find is a poor discoloured worm, as it crawls through
  the mud.
The very sight disgusts us. Only a nameless pity
Stays us from crushing it underfoot . . .'

This from Heine, who must have had his reasons. And of Shakespeare,
no less. As worms go, not the worst of creeps. But in general the cap
fits. Distinctly romantic in manner, of course. Stormy and drangling,
like Hatton Garden on a wet day. What can he mean by nameless pity?
It's a pitiful biographer who fails to come up with a name. Close
quarters . . . and no quarter! Perpetual daylight is our cry.

Prefaces

Was that the primal scene?
It should have been.
And yet he was so young then.
Scene, obscene, unseen?

How could he see a scene?
He had no eyes then.
Nine months to go. But even
So, all the best babes have seen.

Would he see the final scene?
Let it not be obscene.
When what must be would be,
Not much to see.

Growing up lonely, a tadpole against the tide,
He could never put his trust in mass emotion.
Such was his horror of the mob, or else his pride,
That if he noticed people thronging into heaven
With drums and trumpets and Peter as their guide,
He would run at once in the other direction.

   As the twig is bent
   So is the tree inclined
   Bent so often, so many inclinations
   Could hardly see the tree for the woulds
   The murky mights, the tiny maybes
   A laurel bough endures the usual charring
   Branches get cut that sometimes grew.

While waiting for a ship to Egypt –
Filling in at a private establishment
For little misfits from the middle classes,
The slightly backward, the rather delicate.
They seemed to him nice ordinary kiddies.

Mrs Headmaster had her personal family.
He often met her at the foot of the stair,
A stately matron bearing a brimming chamber:
'Milk and honey!' she pealed out proudly.
There was a power of rude health up there.

Noises came through the gilded ceiling
Of furniture breaking, of angers and hungers,
The cublike romping of upper-class infants –
While he and his tender flock sat trembling.

Great debate

Swift lamented that Celia, Celia, Celia ----- [rhyming with *spits*],
but along came Lawrence, crying: ridiculous! monstrous! of course
Celia [rhymes with *spits*], and just as well, it being her proper natural
function! And so battle was joined. To *spit* or not to *spit* was the
burning question. (Not merely when or where.) Some said: great!
beautiful! the more *spitting* the merrier, it was great Nature! Others
felt: but Nature could have been nicer and spared the already humbled
from crouching even lower. It was the hearties and the aesthetes, the
physicals and the metaphysicals, all over again, but more so. This was
the great debate. It still rages, though not always stated in these
particular terms.

Explanations

His folk were excessively averse to exaggeration
And other extremes. They said what they meant.
(Which might imply a poverty of ideation.)
In their opinion, tragedy was best evaded;
They lacked the resources. Comedy had its place,
When it wasn't out of it. Mere misery degraded –
By semantic shift they talked about the weather.

Hardly an auspicious kick-off for an author!
It explains a lot, not least his exaggerated fear
Of exaggeration. And an unprofessional *pudeur*.
(He was rarely heard to talk about the weather.)
His people favoured short sentences and the wisdom
Of simple facts. If depths there really were,
Depths weren't put there for them to fathom.

No wonder that foreigners, the most outlandish,
Were the only subject he felt at home with –
As if they'd want a literature, and one in English!

Absit omen

In Egypt, his little world reeling,
He was sitting and typing his thesis,
'The Search for God'. His skin was peeling.
Scratch, scratch. Drifting away in pieces.

He noticed the flakes were vanishing,
Slowly but surely under the floorboards!
The search was over. God was ravishing
Him shred by shred, dragging him downwards.

Ants it was, bearing away every feather
Of skin for nests, for nutrition.
Symbols were drifting about him, as ever –
A circumstance he was loath to mention.

Catch

Slow to destroy life whatever form it shows
Except for cockroaches, fleas and mosquitoes –
We know the hell reserved for such as these:
Beset by mosquitoes, cockroaches and fleas.

Stones

When a taxi-driver threatens to summon
The Moral Police, that new arm of the law,
And accuse you of criminal conversation
In the back with an innocent Egyptian whore –
Then, albeit at the time you were quite alone,
Be prompt to pay him whatever he asks for.
Foreigners shouldn't cast the first stone.

*

There were sermons in stones,
He remembered, as one whizzed past his nose,

Just missing the policeman who had nabbed him.
The populace were on their toes:
Foreigners were spies, as well as improper,
And some were Jews.
The policeman soon decided to let him go –
Perhaps after all his papers were in order
(Being illiterate, how could the man know?),
Loitering though he was, or moving furtively fast,
In the vicinity of a blatant anti-aircraft gun.
Anybody can get hit when stones are being cast.
You can never find a taxi when you want one.

Appearances

Like high priests disguised as shepherds, there they were, tall figures in immaculate robes, bearing staffs

They sat in the harbour bars, undrinking, unspeaking, like pharaohs disguised as high priests

Wherever he went they were there, imposing, impassive, like archangels disguised as pharaohs

Coffees were hastily ordered, a hush fell, they sat there in state, like gods disguised as archangels

A friendly drunk avouched in a whisper: their staffs were loaded, they were secret police in disguise

The bars were sweetness and light, as there they tarried, like hefty humans with knobsticks disguised as gods

Till they decided at last that he wasn't a Zionist agent disguised as a teacher of English

And they faded away from men's eyes, like guardian angels disguised as optical illusions

Then petty crime revived in the bars, disguise was discarded, unfriendly drunks threatened to knife him.

Innocence

Days of innocence! A knob of hashish nestling
In a mound of damp tobacco. An urchin
Running with relays of red-hot charcoal.
What giggles! Good for the lungs it was,
Sucking at the hookah; if you had good lungs.
They passed the tube around, from mouth to mouth,
Difference of race or caste no more they knew.

Grass was a green stuff then, in short supply.
Cannabis? Something to do with the old embalmers?
Hemp you wouldn't mention in the hanged man's house.
Marijuana could only be a daring Spanish dancer.
(Just to kiss a girl you'd have to marry her.)

Once a windfall: broken biscuits of the stuff
Found in the bottom of a boat that plied the Nile;
No doubt the boatman had been smuggling it.
Then as the sun sank in its wonted splendour
How high they got! Later it proved to be caked mud
Dropped from the boatman's toes. Fine hashshashin!

And now? One at least worn out or dead in gaols;
A fat effendi sweating over files; a cotton broker;
A colonel long cashiered; a weary village teacher;
The clever one in Paris, still a starving artist;
And one remembering those days of innocence.

Question of degree

So they fled from Egypt, with the child in the womb, who was safely
delivered in Rowley Regis, not far from Birmingham University. Soon
after there came HM Inspector, asking about his doctorate. An
Egyptian degree? Was it recognized by the Ministry? He didn't
know? He hadn't bothered to find out? Tut tut! But no more was
heard of this delicate question. The Ministry didn't know either.

Circumstance

Adult classes in Cradley Heath were full of characters;
They had been together for ages, they never grew tired of
    themselves.
Among them, a retired schoolmarm, a bit of a Buddhist,
Who doted on poetry and found the others unrefined;
An old-time chainmaker referring everything to Marx;
A knowing little grocer spotting sex behind each scene
(Such disappointment when they came to *Sons and Lovers*);
A black-browed preacher who allowed of only one good Book;
A motherly housewife representing wives and mothers.
You always knew where they stood, they all had circumstance pinned
    to the ground.

He would think of his other students, in Egypt –
Sent down to prison for rioting out of season,
Or summoned back to villages to feed their families.
Several went crazy, one of them during a lecture:
Was he berating Britain? No, the others assured him,
No need to worry, he was only thinking he was Allah.
A few rich boys waiting to inherit and abscond to Paris;
And the shy bright girls, bunched together in the front,
Touching their cheeks to check they'd left their veils off.
You never knew where they were, circumstance had them all by the
    throat.

Animal lover

Once, something tells them, it was all theirs.
A fair amount still is.

The burden lifted from them,
These days they pledge themselves to leisure,
Cultivate their tails in others' gardens –
Quondam statesmen, finicky about food,
Twitching their whiskers at birds in bushes.

Poetry shines from behind their eyes,
But they will not read it to us,
Content that we know what they think of us
(And still admire them),
Famous for certain though unspecific wisdom.

Their successors have failed to rid the world
Of dogs; and even regard them as friends.
Ah well, that regime will soon be over too;
Surprising they ever got themselves elected.
They too will turn into predecessors –

And remembered for some mysterious wisdom?
A cat yawns sceptically: Wouldn't bet on it.

The pattern was set by a young tomcat called Tom who turned up at
the door one evening. Tom was a fighter, and by the looks of it Tom
was a loser. Tom never won fair lady. He developed abscesses and sat
gloomily on the mat while they dribbled like tired volcanoes. He was
love's victim.

One way to stop this, said the vet as he reached for his scissors. Poor
Tom would nevermore desire to win fair lady. He put on weight and
muscle. His motives were unmixed, his claws pure, it was the Way of
the Samurai. He tore his old rivals apart – abscesses were for others –
and chased off the neighbouring boxer, slash! across its broken nose.
He gained the respect of all. Fair ladies fell at his feet and were
spurned. Tom became a power in the alleys. His motto was: With our
stripes are we healed.

First impressions

At Tokyo airport, descended from a Comet,
He is asked what he thinks of Japan.
The only thing he can think is – Japan
Is agog to know what he thinks of it.

These days he wouldn't dare to ask
What they might think of Britain.
And these days would they bother to ask
What he thought, if he thought, of Japan?

Going cheap

There are some will be heard on Judgement Day
Wailing that their souls have been underpriced.

Invited to do a little informal informing
On the politics of Japanese intellectuals
('To you they'll talk freely'),
What stuck in the throat was the rate for the job.

An occasional trip to the great metropolis
With bed and breakfast at a three-star hotel –
A real treat! All paid for by the Embassy.

Do not litter

A day trip to the sacred island
Where birth and death are both illegal –
Men throwing out their chests and hooting lustily,
Women pulling stomachs in and looking virginal,
Boats on the beaches waiting eagerly
To ferry suspects to the mainland.
Hiroshima is the nearest city.

No show

Stripped to the buff the Japanese girls trip on –
One of them wrestling with a torpid python –
Slowly slowly *dan-dan* they put their clothes on –
Even the python stirs – oh the mounting tension!

Entertaining women

In a night-club in Hiroshima,
A combo playing noisily,
A girl asked sweetly, *Kohi shimaska?*:
Should they make coffee?
No, he replied, it kept him awake.
It was *koi*, it struck him later, not *kohi*:
It was love she had offered, not coffee.
The thought kept him awake.

Next day, as a guest of Rotary,
He conveyed (without authority)
Fraternal greetings from Cradley Heath.
Waiting outside was a victim
(Rotary does not entertain women),
A victim for him to see, to see him.
Him with his face still scarlet,
Her with her white scarred arms.

A child's view

The amah was Mrs Okamoto, a good Christian
And her loving crony. How proud the child was
When Christmas came and her dear one's praises
Were sung by all: 'Okamoto ye faithful!'

They pored over pictures of Hiroshima: ashes,
Shadows of people printed on a bridge, a hole
Burnt in a youngster's pants ('It must have hurt!').
Who was it did those bad unchristian things? –
The Aremicans it was, the cruel Aremicans.

There were foreigners scattered here and there –
A few Eikokujin and the indistinguishable
Amerikajin, and the more exotic Furansujin
Like her mother, some Orandajin, an old Doitsujin
Teaching German at her father's college.

But none, not one, of those heathen Aremicans
Who burnt holes in you and turned you into shadows.
Just thinking made you shiver. But not too much.
They were far away, you would never meet them.

Refinements

'There's a lot of loose talk about the *feudal society* of the bad old days,'
complained the Prime Minister's brilliant son. 'Also about the miserable condition of women. Let me tell you a story.' And he did.

Of how in those old days a gentlewoman, travelling in palanquin or
carriage, might experience certain known though unadvertised urges.
Whereupon, feeling faint as is a lady's prerogative, she would halt at a
country inn and ask to lie down for a while. Bowing low, the innkeeper
would reply that his wretched establishment was indeed honoured:
pray grant him a few moments to prepare a room.

She lay on a thin pallet, taking her ease. Underneath (perhaps less at
ease) lay a sturdy village lad, speedily procured and primed. Neither
took note of the other, despite the curious apertures in the pallet.
What is not observed is not official, what is not spoken of has not
happened. Eventually she would rise, feeling refreshed, and thank the
innkeeper for his hospitality, while pressing into his reluctant hand a
number of coins. Some of which, we trust, found their way to the
village lad, by now back in the fields.

'Such refinements are the mark of a superior civilization,' the Prime
Minister's son declared. 'So let's hear no more about that stuffy and
repressive *feudal society*.'

Samurai

Those squat fellows, like worried gangsters –
Was it the presents they were guarding?
No, they were keeping out the bridegroom's
Army of strange friends and queer admirers.
You wouldn't want too many at a famous wedding.

Aesthete or soldier, it's sincerity that matters.
Later he sought to raise a Patriotic Army,
But the troops preferred a quiet life.
They didn't approve of imaginative writers
Shouting out eccentric orders.

So he slit his belly, to show them,
Crying at the last, 'Long live the Emperor!'
Something they would rather not have seen.
At that time such courage was in error.
How embarrassed the Emperor must have been!

At the theatre

With his faithful follower Benkei and a few retainers, Yoshitsune
approaches the barrier at Ataka. It is guarded by soldiers of his jealous
brother, the Shōgun, who is hounding him. The fugitives are disguised as mendicant monks, except for Yoshitsune, who is clad as a
coolie. 'If you are monks,' says Togashi, the captain of the barrier,
'you must be collecting for some temple. Pray read out the subscrip-
tion list.' Benkei pulls out a paper and pretends to read from it.
Togashi sees through the ruse. He is a man of honour. So are they all.
Togashi makes a donation, and allows them to proceed. One of his
guards recognizes Yoshitsune: 'Stop!' Benkei hurries back. 'What a
red-letter day for this wretched coolie – to be mistaken for the brother
of the Shōgun!' He thrashes him lustily for lagging behind. 'I shall
beat him to death to allay your suspicions!' What agony for Benkei to
strike his Lord! What agony for Togashi to watch Benkei striking his
Lord! What delicious anguish for the audience to witness Togashi
suffering for Benkei, Benkei suffering for his Lord, and his Lord
suffering! Men of honour, all. Togashi lets the party through. Benkei
prostrates himself before Yoshitsune and weeps. A song is heard: 'On
mountain tops have we slept, in open fields, at the edge of the sea . . .'
Togashi reappears and invites Benkei to drink with him. Benkei may
not refuse . . .

They dip into their boxes of rice and fish and pickles, into their bottles. Tangerine peel flows round them. It lasts five hours. Emotions make you hungry. The connoisseurs nod lightly or smile tightly. Others roar applause. Women weep: 'Oh the pity!' Children stumble off to the *benjo*. An actor freezes into an extravagant pose, a stylized grimace. 'That's it!' they shout. 'That's what we've been waiting for!' They know a famous moment when they see one. Benkei entertains Togashi with stories, he dances, pretends to be drunk. 'Still thirsty?' someone jumps up and yells, 'I've got some saké left!' Wooden clappers signal the climax. (And wake a baby up: out comes an absent-minded breast.) To the sound of a drum, Benkei begins his triumphant exit, along the 'flower way' through the audience, taking his time, in ecstatic monstrous hops. An old man calls out the actor's name: 'You're every bit as good as your father was!' . . . A loudspeaker thanks them for coming, reminds them not to forget their belongings. Sighs are sighed, eyes dried, breasts tucked away, feet shod, legs stretched. Coats, umbrellas, babies, lunch-boxes are remembered. A great and good time has been had.

In his opinion, that was *real* culture, something always being talked about but, in his experience, terribly rare. He tried to remember his belongings.

Old legend

It happened that an ogre came among them
Attended by a female and a child of the race.
Child ogres are often winsome, as yet unspoilt;
The grown female defies classification.

The ogre belonged to the *sensei* clan
Which was reckoned to possess obscure powers –
At that period virtually every prefecture
Considered it politic to have its own ogre.

This ogre was a mighty eater of raw fish,
A ravager of inoffensive haikus,
Enemy of the *ofuro*, decrier of kimonos,
A constant complainer of cruelty to cats
(Yet he sought to gas their young in leaky ovens).
He uttered muddled speeches on Important Ogres
Of Past and Present, their Lives and Letters,
And issued proclamations concerning the delights
Of certain shady districts which the natives knew of
But did not wish to hear about.
A long nose, as the proverb says, is sure to poke.

The ogre was tireless, embarrassing the guardians
Of holy places with his would-be compliments.
Inebriated when all around were cold sober,
He offered to entertain the provincial gentry
With rude dances peculiar to his sort.
Among the festive, he sat beneath the cherry trees
Mumbling and shaking his head.
Often he approached virgins of good family, asking
If poverty had obliged their parents to sell them.
Rumour had it he was compiling a Chronicle
To explain the natives. This caused amusement
For he was totally ignorant of their metaphysics.

He meant well, they knew, he was an earnest ogre.
At last he departed. A small crowd escorted him
To the waterside. There he was heard to say:
'I am distressed by the weight of my sins –
Vouchsafe me a sutra.'
Or was it a suitcase? Legends are wrapped in mist;
Police records of the time are not reliable.

From the ashes

In new Berlin they open lesbian bars
Where new Brünnhildes smooth each other's hair,
And burn up Camels. They won't sell themselves
For all the fine tobacco in Virginia.

Whoever didn't do those dreadful things
(Ruins where decent people once read Rilke)
Have had some dreadful thing committed on them
(Darkened workshops of the late persuaders).

All ways it breeds a dour uncertain temper,
In prisoner and guard, in raped and rapist.
Also a weighty pride in stainless towers,
Bright highways, cheaper smokes, and opera.

We all adapt. Always there are bigger villains,
Other victims, somewhere. There are always
New theatres, new Rilkes, new persuasions.
Life must go on, with so much gone already.

Without end

Close ties between old enemies,
Swapping stories of lost sons or comrades
At cocktail parties. Old soldiers
In one another's arms.

(Flags in the breeze; Spandau,
Heerstrasse, Brandenburger Tor.)

Chilly towards old neutrals –
But how they detest old allies,
Devious and unloving,
Who cost them sons or comrades.

Sunny Siam

The land was famous as a land of smiles
('He who doesn't smile is ill,' they said),
Its calmness ruffled by occasional coups
So brief and cliquish as to pass unnoticed,
Except a corpse or two uncovered later
(Never deep down: digging was uphill work
For people not exactly famous for it).
Even the corpses wore a tortured grin.

Night Life

The busy bells of the trishaws
Chirp in the velvety night.
Here is a house of little fame,
Blue movies in black and white.

A woman attends the projector,
Ancient and prone to fail.
A baby is slung about her . . .
It's hard to make head or tail –

Surely the film is upside-down?
But no one dares to grouse.
Her husband is a policeman,
This is a private house.

All of a sudden the sound comes on,
The first sign of zest –
The baby is taking its pleasure
Sucking its mother's breast.

Desiring a beer

They sat outside in the garden, in a shady spot. You could just sit there quietly and have a drink. The madam brought them a girl. They didn't want a girl, they only wanted a cooling drink, they had driven a long way. 'Rather too old,' they said, being British and polite to the natives, 'if one may say so.' The madam returned with other girls. 'Still too old, really,' they shrugged disarmingly. Honour was at stake: at last the madam swept back in triumph. 'Cannot say too old, this one, eh?' They couldn't. Being British, they rose tremulously, thanked her curtly, left quickly.

Cf. Cocteau

Cocteau reckoned that opium 'chastens our ambitions';
Or else it lends a hand in making good the nicer ones.

One evening, two rowdies invaded the inferior shed
Where aching trishaw drivers smoke the dross of others:
A vulgar brawl, knives flashing, then one fell dead.
These were not patrons, quite obviously non-smokers.

'Lacking opium, a sinister room is merely sinister.'
But once the police had gone, deducing drink and passion,
All turned back to their pipes and their wishes – for
Calm, and clarity with calm, and absence of ambition.

The god faded

For a while, opium seemed the answer.
Peace, affection, and sharpened senses.

Was it poppycock? Opium smokers
Cadged fags on the side, forgot to wash,
Their flesh fell off, chests caved in,
Some would kill for the price of a pipe,
Senses sharpened to the coins in a pocket.

Also recommended are a full stomach,
Regular employment and even a good nature.

Pure accident

Roughed up by a dozen tipsy and excited policemen,
He found solace in sensing, between buffets, certain
Gradations of violence, even faint reluctances
(For whatever reason), and a pulling of punches.
He was comforted to note, between unfatal kicks,
A gun drawn but knocked by someone to the pavement.
All this arose from a pure and simple misjudgement –
In a time between Terrors, nothing to do with politics.

Even so

Even so. A house, a garden, near a river,
Little rivers in the garden. Cool of dusk.
Lamps flickering, mosquito coils, the scent of
Jasmine. A glass or two of Singha. Cigarettes
Like fireflies. Faint gleam of shirt and sarong.
Silent uproar of the crickets. Unweighty words,
Mostly unheard. Two silences. One peace.

But these words dim it. Are there things that
Can't be written, then? Allow a few. Repose.
It's gone, it's almost gone. And even so,
A house, a garden, water, silence, cool of dusk.

Sweet smell

At Bangkok airport, tickled by garlands of jasmine
From students and anonymous sympathizers,
He stands in a daze at the foot of the gangway
While smiling stewardesses interfere with him –
'Sorry, but we have to strip you of your flowers,
The scent's so heavy, the captain might pass out!'

They didn't last for long, those wreaths of laurel.
(He wonders what will happen to the booty,
Someone went to great pains in the weaving.)
Dragging up the ramp, his shirt smelling sweetly,
Puzzling how a poem could upset the ruling Marshal,
He hopes for a captain with a stronger stomach next time.

Singaporeans

Overseas Chinese are the easiest people to live among
They rob or assault or cheat only their fellow Chinese
And only when truly unavoidable or rewarding
They kidnap only the children of their fellow Chinese
(Unsure that foreigners care to recover their offspring)
They do not mind in the least what one may say
About the Malays and the Indians and the Eurasians
They do not much mind what one says about the Chinese
Since only the Chinese are to be taken seriously
Or more exactly those from the same village in China –
Indeed they are a most amenable people to live among
Not feeling obliged to have feelings about all and sundry.

In hot water

To debunk the borrowed robes of nationalism
He invented the term 'sarong culture' . . .
It was confused with the popular saying:

'When durians [*highly prized fruit*] come down
[*from tree*] sarongs go up [*supposedly referring
to* (a) *aphrodisiac properties of fruit, or* (b)
pawning of garments in order to purchase fruit]',

And hence taken to insinuate laxity both sexual
And financial among a section of the community.

The ruling party was not pleased with him at all.
'If he were Professor of Sociology, I could
Understand,' said the Prime Minister. 'But Literature –
What has that to do with real life?'

Interrupting the programme

An odd sensation, to enter a favourite bar
And hear oneself denounced on the radio.

(A change from queuing in some homely pub
To buy a pint, inaudible, ignored.)

The waiters were incurious. They had heard
So many scoldings, in so many tongues.

Moon-faced and beaming, to and fro they slid
With trays of drinks, not a drop slopped over –

Knowing that music was soon to follow,
Noble and wicked sentiments from Chinese opera.

Meat counter

'We're on your husband's side, mem!'
Thus the cheery young blades on the Meat Counter
In the supermarket. Chinese of course.
They chopped straight through a side of mutton
To show their sympathies.
He hoped it wouldn't come to that –
Frivolous as ever, drifting over to Wines and Spirits.
How noble the Scotch looked, how innocent the Gin!

Home

The old house was thought to be haunted.
The Chinese avoided it at nights –
During the occupation, or so some said,
The *kempeitai* used it for interrogations;
According to others, it served as a brothel
For Japanese officers.

There were fearful screams, and thuds;
There were grunts, and what sounded like
Rude guffaws and nervous giggles.

– A mass invasion of monkeys,
Bored with the nearby Botanic Gardens
And looking for new haunts.
They swarmed into the lavatory, tugged
At the chain, occupied the wooden seat.
They were a lively lot, they would go far!

One swung on to the power line,
Bit through the insulator, and sizzled.
They had discovered electricity.

And garden

What is more tonic, at times, than the jungle
Just held at arm's length, almost at your door?
Our respects, great Nature, man's best neighbour –
How you harbour chaos and yet preserve the law!
A grave macaque bent double like a gardener,
Raging bougainvillaea, thick-lipped frangipani,
A dewlapped lizard glides from branch to branch,
Or butterflies, those 'flowers returning to the tree',
Crickets and croakers, a snake or two in the grass,
Starling jostles bat till room is found for both,
And orioles of gold, and humming-birds like bees.
Ah the sweet smell of damp decay and sweaty growth!
Nothing can kill this world, you murmur as you ease
Yourself in evening cool, not even us, not wholly.
But don't forget, dilute your urine for the orchids.

Drinking with a Minister

After a ticking-off of dons by the Premier over dinner,
Drinking with the Minister of Health in the old Government
   House –
Everything was about to be forgiven, at least forgotten,
As the brandy went down and a new bottle was called for.
But the servants, a judicious mix of Chinese and Malays and Indians,
Turned sullen: they wanted to close up and go home.
'It is grown late,' they grumbled, adding a sardonic 'Master',
While rattling brass trays and switching lights off.
They had borne with the funny ways of the British,
From their own they expected more consideration.

Rest and Recreation

Why, it was almost Princeton! The GIs
Scattered cautiously across the campus,
Casing the idle buildings,
Leaving notes in pigeon-holes:
'Prof, I wanted to audit your lecture
But we go back early Monday.'
In Vietnam they'd forgotten about weekends.

The rival professor

Then a rival professor came on the scene. An American, sophisticated
and *à la page*, an expert in linguistics. (So much more apropos than
value-laden literature.) He was entrusted with a delicate official
mission: the tutoring of young diplomats from Peking.

This put our man's nose out of joint. Not merely was he on the wrong
side, but now it seemed there was a right side, and someone on it.

Abruptly the rival disappeared. His picture was turned to the wall, his
name no longer heard in the halls of power. It was whispered that he
had links with the CIA. (Unlike Wordsworth, Milton etc.)

The CIA was sedulous but prone to accident. (Admiring the PM's
purity, it once offered him money. To the pure a cash offer is a bribe.)
Yet how adroit it was in its choice of language.

Small honour

Things left undone, or done indifferently.
(As even he would come to see.)

Lectures that added blur to blur,
Pepped up with digs at the ministries;
Poems that started out in rage and pity
Then turned into harmless obsequies.
(To go no further.)

Counsel spooned out, the Wisdom of the West,
To keep his end up. Reasonably prudent,
And yet his great success: a criminal operation
Procured in secret for a negligent typist
Pregnant by an indigent student.
Such gratitude! She told the world.
(Later it seems he took against abortion.)

Courage enough, or carelessness, to get him into
Temperate trouble. A deportation order
On the cards. Phone tapped. Book banned.
Shocks that flesh and speech and print are heir to –
For whose good was it? His own small honour?

So long to learn, no time to be a teacher.
Better to give to beggars, no questions asked.
Then offered an OBE for 'services' –
His former pupils staffed the ministries.
(To go no further.)

You're not at home now

The ten commandments of expatriation
As drawn up in one's darker moments;
And later reduced to five, at the darkest
One being neither God nor Moses.

Confine yourself to value-free activities;
Advising the army, treating hernias in high places.
Otherwise keep to your ivory tower,
A contemptible address but unvisited
By the Special Branch.

Do not presume to complain of the heat
Merely because you see a native sweating.

Easy for you to like everybody;
Risky for anybody to like you.
Remember, the two old gods still rule:
Must-do and Make-do.

Judicious sex is permissible;
Sex has always been a substitute for politics.

Do good in secret,
Give to the Cancer Research Fund.
Cancer has always been a substitute for politics.

Your day will come.
City fathers will name a cul-de-sac after you.

Pom

Advance, Australia fair!
And so it did.

They didn't think he was a Pom.
(The Johore Professor of English

Had to come from Johore.)
They thought him 'a Malay of some sort' –

Albeit, like Australia, fair.
So he advanced.

In China

Lots of reproaches in China . . .
For not totting up the Hong Kong coins
In his pocket. And then
For frivolously declaring a trickle of cents.

For occupying a huge hotel room on his own.
And for proposing to share it.

For liking the cheongsam.
For suggesting it was the national dress.
For asking where it had gone.
('Gone to – ha – to decadent countries.')

For neglecting to press a cigarette
On the guide every five minutes.
For pressing a whole packet on him.

For scaring a group of small children
Into tears.
Then for making them giggle.

For being what was expected.
For being unexpected.

For not wanting to visit a collective farm.
For wanting to visit a university.

For mentioning *Sons and Lovers* in the staff room.
For declining to read a poem by Julian Bell
Out of Quiller-Couch's *Oxford Book*
On the grounds that it wasn't in it.

For being cultural instead of linguistic.
(Perhaps they could sense the Revolution.)

For enquiring after the Faculty of Law
('We do not have one! We do not need one!
We have no personal property and no divorce!').
For unlawfully upsetting the Dean.

For failing to laugh aloud at acrobats
Clowning around in US army uniforms.
For showing shock when a spectator hawked noisily,
There being a law against it.

For advising the guide to get some sleep.
For disrespecting the guide's solicitousness.

For ordering a second bottle of beer.
For starting the Opium Wars.
For offering his regrets.

For wishing to be friends.
For suspecting they might like to be friends.
For seeing that nobody could be friends.

He felt guilty if he enjoyed himself.
Enjoyment lay in feeling guilty.
Being scolded made him feel at home.

Did they feel at home, he would wonder,
When the Red Guards arrived soon after?

People's museum

In the museum hung a painting of tiny ships,
Flying huge Union Jacks and confronted by tall
And valorous peasants with cutlasses.

Another picture showed Commissioner Lin
Tendering ritual apology to the Spirit of the Sea
For dumping confiscated opium.

Before our time, the guide may have implied
By his strangled titter. The bemused Britisher
Grunted in muted apology –

Relieved that no one seemed to be denouncing anyone
At the moment; disturbed that so many seemed to be
Looking over their shoulders.

It was peace, for a time, in a place.
But the welding of nations is not so easy:
Old Mao was about to make waves in the Yangtse,
Without a word of apology.

Armies

As if so many princes weren't enough
They fashioned generals, often out of
Sergeants from colonial legions.

'Good iron isn't used for making nails' –
The people's wisdom sometimes fails.
Nobles are born, soldiers are made.

(He'd known a minor noble working in a
Filling station: 'Blame it on polygamy.'
Also a princess who taught philology.)

The women all looked like princesses.
Virgin corporals in battle-dresses,
They learnt the ways of the soldiery.

Dense with relationship, these sudden armies
(Be careful not to rape your sister!).
In extenso died the scattered families.

Still many princes left and many generals,
Removed to milder spots where rank still tells.

Off limits

Ponto-chō, heartland of the geisha, in Kyoto: a narrow street of wooden houses. One evening a jeep drew up at the open entrance to a little bar. Come out, ordered the US military police: Let's see you. To which our hero stoutly replied: You come in here. No, said the MP: It's off limits. (Not that the US Army had declared the bar off limits, but the bar had declared itself off limits to the US Army: there was this reciprocal arrangement.) Come out here, the MP repeated. Said he, insouciant: You can't arrest me, I'm English. A muttering under the breath, and away the jeep bounded, along the bank of the gently flowing Kamo.

But those dusty charming towns of long-gone Indo-China. Shady trees and hotels, peaceable slow-moving people, given to neatness (and to their neat children) and not-too-strenuous pleasures, graceful in their mild disgraces, a civilizing influence. Trishaw drivers chatting in French; now learning English, language of the future, a millennium of opulent fares. You can't arrest me, I speak English.

Process – how cruel it is. Nature abhors the vacuums we create, fills them with more like us. Filled women, cradles filled, streets filling up again. Ponto-chō stands where it did, a nest of singing girls. But gone are old Phnom Penh, Saigon. Come out; off limits. Such vanishings, such bans . . . like age – a kind of house arrest.

Wisps

But these little ones are so very ignorant,
It is hard to know where to begin with them.
Mostly there is something to build on, to put to rights,
Some special and meet desire – how we run to fulfil it! –
Something to complete. Here there is everything to complete.
All they wish for, impossibly, these wisps, is to be no more than they
    were, no other.

They block the roads. What realm are they subject to? –
So skimped, unlicked, such blank unwritten pages.
Folklore tells of a novice in hell, facing examinations,
Who burns exemplary answers and swallows the ashes.
They have swallowed ashes already.

What's to be done? While we ponder (we have to ponder?),
Let them be placed in – to borrow the word – in a camp.
For their own good, surely. (It has to be better than good.)
In a happy land, it goes without saying! Where sun shines softly, rain
    is gentle, peace drops slowly.
It isn't easy, best will in the world, in the best of worlds,
To provide for such guests, for these tiny ghosts.

The real thing

Compassion was a man-made thing,
A product of our affluent decency.

These could not afford it. Something older
Moved them: Stay alive, see to the children.

Heads were kept down; or heads might fall.
(Keep your foreign conscience tucked away.)

In these, compassion was heroic, saintly;
It had to be. Now and then you meet it –

The real and purely natural thing.
A privilege not to be forgotten.

The old days

Whatever the creeds, the policies, the movements,
Behind them straggle the ghosts of starving peasants.

On this earth in the old days there used to be giants.
No one reported the starving of peasants.

For all his fears and flinching, at least he never
Was much of a praiser of times past and over.

Lacuna

So little about the power of religion! And in regions renowned for powerful supramundane faiths. He might say, he saw little of it, it was obscured by the power of secular doctrine. Yet in one country, the student who believed he was Allah. In another, the Director of Textbooks (Min. of Ed.) who was quietly sure he was Buddha. Was he himself ever tempted to think he was Jesus? Not seriously. He recalled that Christ had been crucified. For a religion, mundane, all too mundane.

Spring clean

'If my devils were driven out, my angels too would receive a mild shock.' One takes the German's point, at once both poetic and prudent. But might perhaps the angels, mildly shocked, be less stand-offish then? If one cried, at most they murmured 'There, there', always having something better to do. They were rarely at home; one wasn't really at home with them. Immaterial seemed the word. Once or twice one may have entertained them unawares.

And afterwards, would the devils slink away, tails between their legs? Or merely strut more insolently? He could hear them jeering: 'Give a dog a good name, it'll still bite.' Devils were probably irredeemable.

Yet, oh to fall in love with a beautiful psychiatrist! With grey eyes, long neck, chaste white smock, motherly bosom. A wolf-man he! A ministering angel she! To take her away from it all. Or would she soon grow bored? (The *Standard Edition* of S F reads like a novel – like sf in fact.) Even though he told all. And it didn't do to tell women too much.

Or men, which most of them were. Flat-chested father figures. Or school doctors: 'More outdoor sport for you, young fellow!' Devils v. Angels at basket-ball? How politely Rilke declined the offer of 'a great clearing-up.' Heine would have emptied a chamber-pot over their dingy beards.

Better the devils you know . . . till the last dim beating of that other angel's wings.

The biographical mode

Observing their subjects, biographers grow curious.
Observing biographers, their subjects grow cautious.
The bonfires of the latter make the former truly sore.
Letters can prove a lot, but burning them proves more.

Naming names

A page in cipher –
But somehow sounding female.
Could be rhyming schemes
Possibly pet names of visiting typhoons
Perhaps the flowers in the Garden of Eden
Conceivably the peculiar terms in which a Thai princess addresses a
    Thai princess on the subject of Thai princesses.
(More likely the women he slept with.)

Watching one's words

The word *homosexuality* implies the having of sex.
Henceforth we shall use the expression *homoeroticism*.
Undoubtedly there were men he was very close to.
(Their names will occur no doubt in other contexts.)
He can be said to have liked if not to have loved men.
Hence we prefer to use the expression *homoeroticism*.

The *homosexual* is licit now, and some say even gay.
But what twisted tastes does *homoerotic* conceal?

Short thoughts

Returning to your native shores, what is it
Hits you between the eyes? The amount of dog shit
In evidence. It may be heaven, it may be hell –
One thing's certain: pets are used to eating well.

Pointing to Goethe's works, Freud identified evasion:
  'He used all that as a means of self-concealment.'
What a lot of self to conceal! What frailties unseen! –
A hundred and forty-three volumes in the Sophien edition.
Not much self in Freud, or else a lot of self-revealment:
In the end his *Gesammelte Werke* only comprised eighteen.

          Nature was best left
          To the specialists. It reminded him
          Of cross-country runs at school
          In wind and rain.

          Then there was Love. Severe girls
          At poetry readings chided him
          For not providing it.
          He left it to the specialists.

          From which it's easy to deduce
          A philistine and misanthrope.

They love us so much they could eat us.
Then get together to compare stomach pains.

But for them we would never have come here.
We must pray they never lose their appetite.

          Quite often heard to call on God –
          Though not expecting an answer.

          For who else could he call on –
          On some temporal lord and master?

          Angry with the one for being there;
          With the other for not, still angrier.

'You – the suave tempter promoting
An aperitif! Where's your dignity?
You, so – deep, so – distinguished!
You can't need the money.'

'You ninny – when God does voice-over
In commercials for nappies . . .
Dignity? Need? We move with the times,
The times are our business.'

317

It isn't bad literature we shall die of,
But good television: brilliant pictures that can
Only tell one story. Of what man does to man.
The camera can't lie all the time. Switch us off.

By nature pedantic
and deaf to ceremony
suspecting that agony
was rarely romantic.

The least one can do for one's family
Is dedicate one's books to them;
And the most one could do for them
Would be to swear off writing totally.

There used to be writers and readers
So readers would profit from writers
And writers would profit from readers
But the latter learnt more from the former
Than the former could learn from the latter
So the readers then became writers
And the writers no longer had readers.

He used to hate the laws. He broke as many
As he had the strength to. Or he bent them.

Now others have cast down the laws. He weeps
With rage to see them lie there, broken –

More laws than he had ever heard of,
Lying there, not bent, but broken.

In another time, another place they could have been friends. They had
so much in common. Then and there it was not to be. We all have so
much in common. In another place, another time we should be
friends. Long ago and far away they would have been friends, in an
earthly paradise, we should all have been friends. In a little time, under
the earth, we shall be friends. We have so much in common.

A founder member of the party of one –
Then did he resign or was it expulsion?

At what age was innocence lost?
Not long after birth
It must have been. And then –
An eerie sense of arsy-versy
Or what Shakespeareans called
'Inversion of the natural order' –
Seeming in pennyweights and scruples
Slowly to regain it. By what age?
Perilously near death it must be.

Posterity was always a great reader.
He would beg, borrow or steal books,
He would even buy them.
You could be sure to find Posterity
With his nose in a book.
(Except when listening to music
Or peering at paintings.)
He had excellent judgement too.
You could always put your faith in old Posterity.
We shall miss him.

That long disease, a Life: wind, diarrhoea, constipation, diarrhoea,
earache, toothache, mumps, acne, black eye, common cold, headache,
toothache, chilblains, pink-eye, cysts, prickly heat, chills, quinsy,
arthritis, toothache, piles, headache, arthritis . . .

And he leaves to the members of his family
A Life of their own for each of them.
It is unseemly that guiltless bystanders
Be splashed by badly conducted vehicles.

Fear and knowledge

Living next to an ancient primary school
TO THE GLORY OF GOD
FOR THE BUILDING UP OF THE CHARACTER
IN HIS FEAR AND KNOWLEDGE –
Maybe some of it would spill over, not only
Toffee papers and finger paintings . . .
Here they come. More mothers than children?
No, just bigger. More cars than mothers?
The children scatter from under their feet.
Blonde dolls in peasant skirts and blouses,
Pint-size scruffs in denims, parkas, plimsolls.
Only the gleaming little black and brown boys
In honest shoes, white shirts and proper trousers
And a small trim oriental in (yes) a gymslip
Look at all like British characters.
They shall inherit top marks, for starters.

'Pipe-smoking fuddy-duddies the curse of publishing'

People fond of literature sometimes seek a career in publishing. This
is a mistake, though not always fatal. A poet is creating in Reception: it
is his right to be published, it is your duty to publish him. He is given a
cup of tea. Then comes a person who has discovered the secret of
immortality: it takes an eternity to tell.

Of course one needs living writers, but the dead are nicer. No risk of
Mrs Woolf jumping off a bridge because her new book is full of
misprints. Like the military expert intending to say: the loss of
Vietnam will lead to a blood bath – and in the book it reads bloodless
bath. Anguish. Horror. Commies at the printer's? Proofs are con-
sulted – ah! spacing between words was uneven, and the author asked
for *less* between blood and bath. Oh no, it isn't all bandying philosophy
with Iris Murdoch over scampi and Chablis at Giovanni's.

He had his uses, an almost infallible judge. What he admired wouldn't sell and what he didn't would. Dark doubts assailed him. If you shrink from a tale of necrophilia, does it mean you are a crypto-necrophile? If stories of raped and murdered children fail to charm, that you harbour like urges? And another thing – if he could write books, why couldn't he write a decent blurb? Because blurbs need to be better written than books.

Ah well, if you want to go on liking literature, think of books as potatoes, sold in sacks. You like potatoes, don't you?

Tribute

Caesar turns up each morning, asking breezily:
   'And what are you offering this fine day?'
Well, there's this thing – and that – and also
   (you suppose) the other . . .
The hours go by. Caesar has time to throw away.

During the weekends he's away on his yacht,
   or visiting Glyndebourne.
God might take the opportunity to show his face.
But would you recognize him after all this time?
Best to put a few pennies by, just in case.

Things falling apart

An alarming sickness spread through the world, a plague which some called by an ancient name, individualism. It came between man and man, between man and wife, between man and master. Empires crumbled, Nations were discredited, the Parties fell apart, the Family broke up, Oddfellows grew odder by the day, the Acacia Avenue Tenants Association lasted but a week.

Hordes of individuals roamed the streets, like mad dogs, squealing like scalded cats, 'me, me!', doing their things, often quite nasty. (Had you avoided contagion? At times you had doubts.) Doubts were raging everywhere, by day and by night, and certitudes were not far behind. There was a great hunger and thirst, and little left to satisfy the needy. Some welcomed this course of events, declaring it a Sign of Life; often they retired to write their me-moirs. Others considered it uncivil, and blamed mistaken sages and false faith-healers or soul-doctors.

One alone there was who might stem this furious epidemic. But he had taken down his plate and vanished from the consulting rooms. A brainy though depressing German called Nietzsche insisted he was dead. Possibly he had been the first victim. That would be one of life's little ironies!

Parts of speech

Forever rubbing shoulders with words,
He couldn't always tell one part of speech
From another, especially in the morning.

There, inscribed on his desk in large letters,
Was the verb REJECT.
What a way to start the day!
The grim faces of editors and women
Stared up at him, ticket-collectors too.

His eyes focused. There, incised on the pipe
He reached for to start the day,
Was the word REJECT in large letters.
A noun, pronounced as reejekt.

Never mind, he murmured, you're a good chap
Whatever they say,
And you're MADE IN GREAT BRITAIN.
He walked off to work, cradling his pipe.

Where art thou?

He wrote a children's fantasy
Full of clean fun and unspoken morality,
For little Alice, Alice his ideal reader.

He ran to take her the very first copy.
But Alice was living with some feller,
Alice was on the Pill, Alice was crotchety.

He wrote it pretty swiftly,
It couldn't have been published faster,
But Alice was on the Pill already.

In the old days up at the varsity
They talked a lot about 'maturity' –
Could this be what they were after?

Oppressed poem

Who strained its stresses and reduced it
to mean things Who starved its similes
and flogged its feet Who seduced it
rhyme after rhyme and set the doggerels
on it Who crushed it in cold ironies
Who sentenced its tender participles
to dangle and docked it of curlicues
Who hounded it into barren hyperboles
Who was the pronoun who martyred the muse –
Who shall be barred from polite anthologies.

Written off

Sad child, conceived without your willing,
Product of someone's careless living.

Despised of men, if not rejected,
By your sire's iniquities infected.

Unlucky book, never to reach maturity,
Whitely you lie there, staring through me.

Meeting writers

In unaccustomed comfort
Writers were travelling abroad to meet writers.

They met with experts on writers –
On the Writer in his Time
Or in his Place
On the Writer in Society
Or in Isolation . . .

    The writers did not meet with writers –
    Perhaps they were not nice people to meet
    Perhaps they did not care to meet writers
    Perhaps the writers had unwittingly passed them
        travelling in comfort in the opposite direction?
    Or perhaps they were actually writing.

In any case writers convey an aberrant impression of writers.
Experts are always more trustworthy.

Old reviewer

So long a reviewer, he could talk
With a semblance of sense about most things,
From belles lettres to birth control,
From Tantra to the Toltecs.

It was words did the thinking, he thought,
And one word led to another.
They thought of pleas to the tax man,
Of poems and wills and testaments.

The whole of knowledge resides in words,
And words reside in the dictionary.
If someone stole his *OED* they stole his soul.
He slept with Roget underneath his pillow.

But he didn't sleep.
He thought of words like slumber, snooze, siesta,
Hibernation, aestivation, catnap, forty winks.

Cross words

Significant form: shapes of bottles, women, coffins.
Praxis: as long as it makes perfect.
Semiotics: more signs and portents!
Discourse: emetic to end a Chinese banquet?
Structuralism: you need a floor under your feet.
Hermeneutics: an interpreter is called for.
Deconstruction: books remaindered, books burned.
Tel Quel: but will they listen?
The Death of the Author: alas, an eternal verity.

Self-opinion

He might be thought to have thought himself
A middling poet because he came from a midland town;
Or provincial in that he had lived in Settsu Province
(Itself in the Kinki District, which was wasted on him);
Or Alexandrian, or (if hardly daily) Teutonic Knightly;
Or, having worked in the Black Country, demotic and smutty;
Or lionlike in coming into Singapore (and out like a lamb);
Or capital since (with the help of it) he moved to London,
But south of the river, so a simple suburban soul,
He might be thought to have thought himself.

Ambitions

What he would have wished . . . a dozen surreptitious ambitions . . .

To have grown the first potato

To have invented one or more of the following: anaesthetics, the flush toilet, the Golden Mean

To have uninvented no end of things, including the end

Alternatively, to have composed requiem masses noisy enough to waken the dead

To have written like George Herbert (and lived longer)

Or to have written (parts of) the Bible

To have founded the NSPCC, the RSPCA

To have been a hero (without actually killing anybody)

To have been boss (short of bossing people about)

To have been Pinkerton and changed the story of Chōcho-san

To have possessed this man's art and that man's scope (without needing to be either of them)

To have been himself, with certain improvements.

Great chain

The emissaries of Venus
Run riot among the guardians of culture
(Male and female are they also)

Like a tracer, the twisting streptococcus
Plots the intellectual luminaries
Like a string of bulbs on a Christmas tree

Like a radioactive isotope
The holy spirochaete
Links the media in spirals of loving

Who reviewed whose book and the wages thereof –
For the eye is not satisfied with seeing
Nor the ear with hearing

And would there be peace now
While the high ones lick their wounds
Would the thunder of brains desist
While physicians attend the lower regions

No, for the clever are not like us
Pain is a spur that the clear spirit doth raise
To lead laborious nights and days –
Here come more doses of great fame.

Sadder, the massacre of an innocent, a symbolic event. Poul was an
artless young Dane, melancholy, sensitive, shy. Never had he heard
those bold words: 'On our sleeves we wear our sexes,/Our diseases,
unashamed.' In Krung Thep (we prefer the official name on this
occasion) they always asked him to their parties. Too shy to speak to
anyone, hardly drinking – but he liked to come, he said, he would just
stand and look and listen. Would he had done no more.

Poul disappeared. Had he gone home on leave, too modest to mention
the fact? They discovered him in hospital, a case of advanced syphilis.
Resolving to wean himself away from shyness, he had gone with a
woman. Too shy, too modest, to look elsewhere, he went with a
prostitute. Too shy to go to a doctor. It was his first woman. His first
job too. Shocked, alarmed, the East Asiatic Company sacked him.
Such can be the wages of shyness, sensitiveness, sex.

Fathers

Visiting Belfast, he would hardly have boasted
Of how his father boasted of a Fenian father.
Remembering rather that many years before
His father was taxed with the small explosives
Stuffed into pillar-boxes –
The boxes that it was his father's job to empty.

Small cultural objects

'America, you have it better,' Goethe observed
Plaintively. Certainly better than old Poland.
For instance, the carefully faded elegance
Of the Hotel Europejski, once the pride of Europe,
Its marble stairs redeemed by a ragged carpet.
In the foyer, behind large slices of plate glass,
Small cultural objects cower, for sale or not,
And a native poet confides to a foreign one
That three poems have been banned from his new book –
Should he withdraw it? Or is half a loaf better . . .?
An enormous radio is transfixed to one transmitter:
All the better to be heard. Outside the chamber
Sits a deadpan chambermaid. Doubtfully for sale.
Servants are not servants, they are civil servants,
They wish to meet your wishes, which had best be modest.
Back in the far-flung shadows of the restaurant
A cluster of rustic hussies try to remember
What cocottes were like. The Palm Court players
Explode in a waggish folksong, and they stop trying.
Waiters whisper hungrily, something about dollars.
Are they provocative agents, like the hussies?
Here are the grand menus still (stained with bortsch),
But not the dishes. Half a loaf . . .

The elusive sage

The so-called 'Sayings' occupy a thick exercise book, but for present purposes a handful of examples will suffice. To begin with, there is an element of mystery attaching to the urbane Chinaman whom he claimed as a lifelong friend and mentor. Sinologists are of the opinion that Tao Tschung Yu probably lived (if at all) in the eighteenth century. Many of the dicta support this view.

'The Emperor speaks proudly of the young men he sent to their deaths; in the palace window a candle weeps for a straying prince.'

'Who sorrows for the long nails of the mandarin? Only his wife with the bound feet.'

'"Too thin-skinned for words!" complained the cow as she rolled over on her friend the butterfly.'

'Poetry teaches the correct names of birds, beasts, herbs, trees, persons and places. Do you ask for more?'

'He who makes an offering as the temple burns down is an idiot; or conceivably a superior person, psychologically speaking.'

But was there psychology in the eighteenth century? Possibly the Chinese invented it. Many of the 'Sayings' strike us as curiously modern in tone.

'The secret of an orderly and stable regime: writers should be obscene and not heard.'

'The Minister for Culture wants one urgently? Place some agar-agar in a saucer . . .'

> 'Politics has overrun our beds – the last border!
> It lusts for us, the thing we lusted after.'

'People may be divided into two classes: those who think that if weapons of war abound they will not be used, and those who think the contrary. A third and more numerous class prefer not to think.'

'With so much violence around we shall soon see a war.
With so much violence around we do not need a war.
With so much violence around we have a war.'

'The work of the devil who rules over printing-houses: *world*
supplanted by *word*.'

> 'They juggle with some great abstraction,
> And study the small print.
> Their love for man is beyond calculation,
> They keep a numbered account.'

'Like mothers, critics wish you wholly other than you are. But
they do not feed and clothe you.'

'A dream: the awarding of bursaries to non-writers to enable
them to continue not-writing.'

> 'Line after line incised: that's character!
> Then the knife slips. Back to the maker.'

'That great guy Paterfamilias – saint? philosopher? dreamer?
brute beast? Whichever, let us honour him.'

The sage can occasionally sink into facetiousness, e.g.

'Who are the most illustrious of the people of Ch'in? Fu Manchu
and Confucius He Say',

and although the following aphorism is enigmatic in the Eastern
fashion,

'As the noun declines, so the verb fails to conjugate',

scholars point out that Chinese is uninflected, and hence the language
does not lend itself to the putative metaphor.

Did he really have a friend called Tao Tschung Yu? Or was this a man
of straw, a stalking-horse, a mouthpiece for dubious wisdom? The
final entry may conceal a clue.

> 'Avoid him who forgives himself everything.
> Avoid him who forgives himself nothing.
> Cherish him who even forgives this moralizing.'

Sleeper, awake

A poet told of enjoying in a dream the ugliest and
dirtiest girl imaginable, when for the same price
he could have had the most beautiful princess . . .
In our dreams we are often surprisingly modest.

In his dreams he was customarily fairly modest,
apart from catching the youthful Princess Elizabeth
when she tumbled from what seemed a circus elephant,
and once composing a late-Jamesian novella . . .

So he knew he was sick, when instead of missed trains
or friends, or missing lecture notes or trousers,
he dreamt of prodigious signals flashing in the sky,
and singular emotions, and the world's end imminent.

Even then not much happened. At the eleventh hour
he woke in a sweat, and found it well past midnight.

Fears

It couldn't always end in clownish despair.
There must have been real depths somewhere,
And even relative heights. He must have won
Something. Sometimes have heard *koi* rightly.
More fearful he was than an ancient Briton –
Wood touched, fingers crossed, lips sealed tightly.
Evil had eyes. Hate lay in wait for love.
Fatima's hand fitted him like an old glove.

What you love, never like too much. That which is given can be taken
away. He jests at crossed fingers who never felt a loss. Lonely the glove
whose mate is mislaid. Idle talk can cost the lives you love. Drowning
in depths, falling from heights – man is a reed that barely dares to
think . . . These are our homeopathic bromides, not to be scoffed at.

As for 'hearing rightly': maybe, but a true story lasts longer than a
sham lay.

Love poem

'I did not promise you a rose-garden
I did not mention an epiphany
(Yet you are the wisest of readers)

Nor, although pushed, has language fallen
Never did sin come into the word
(Even if innocence has long departed)

I merely offer whips and guiles
Cods and wrecks and wreaths and biles
Roughage by which our minds are moved
Fallen or sitting or rising up

Gladly would I give you a rose-garden
If life were a bed or I had green fingers
If I were Christ, I would think quite seriously
Of arranging a grand epiphany.'

(Though described in a note as 'about the fall of language etc.', this must surely be one of his rare love poems, and it is printed here as such.)

Like mother . . .

Click, click! She adding to the available stock of cardigans
– For his birthday she might give him a good grey pullover –
And he, though less equably, to the stock of phrases.
Busy hands, each carefully tolerant of the other's fashions
– Once he dedicated a nervous book of verses to her –
While wondering who would want the words, who wear the
  woollies.

Dropping names

A pusillanimous view! – that you cannot decently write about the living unless they are unrecognizable, or illiterate, or live at a safe distance and will never know, or are as near death as makes no difference. (Or, very occasionally, are indubitable and unmitigated scoundrels; or established saints.) Or of course – as his mother became – they are virtually blind. (What she most missed was knitting, and television, specifically the advertisements, which she found more consistently agreeable than the rest of it.) Since self-denial of this kind affects the most interesting things in life, such *mauvaise honte* is a great shame.

Around the Styx

Nothing of a Berryman,
Closer to a ferryman –

Hanging around the Styx
For evidence of dirty tricks

Where come the dead and dying
With their unregarded crying

(Charon describes his passengers
As mostly innocent bystanders) –

Or, piling man-made miseries
On natural calamities,

Typhoon and earthquake, fire and flood,
Costly ideologies, cheap flesh and blood,

More like Old Moore's Almanac
Engrossed in universal wrack.

Untold

Memory, our nurse and friend and patron –
Until by freakish maladministration

She summons up those ghosts, both new and old,
Kind and unkind, whose names we leave untold

For fear of giving flesh and blood and bone
To what we then must own to or disown.

No end in sight. They spread from coast to coast.
Once heaven or hell was where we sent each ghost.

Behind, in front

He had long felt himself implicated in that
'terrible aboriginal calamity' noted by Newman;
whatever its purposes, creation was out of joint.

Curious then that with the years there grew a sense
that this calamity was yet to come, this huge mishap
of purpose, in some ever-approaching future;

no abstract primeval perversion incurred out there
among the infant planets, but not so far from here
and now, the putting out of joint of him.

Actuarial

They're all agog to handle your cash
Even though you have none –
But though you still have one
No one wants to underwrite your life;

Only your roof, your bed, your desk,
That faithful if unsteady clan.
Approach the Capital Builder Plan
And you'll be shown the door.

(The cult of youth was founded
Soon after he turned forty.
Soon after he has ceased to be
They'll introduce eternal life.)

When things seem rather safer
Than they often were, it's sad
That you've become so plainly bad
A risk. Well then, be bad, take risks.

Auguries of sinfulness

Be bad? Less easily done than said
(and if done, not to be read).
What is the wickedness reserved for age –
to put the National Health in a rage?
Arson's as out of place as Sodom;
the poor-box electronically secured;
language already well manured.
When the blood runs sluggishly
conscience can catch up quickly.
Seriousness seeps in,
like syphilis (*v. supra*) . . .
One is old enough to be anybody's grandpa.
(Maybe they do it different ways
these days.)
After four glasses, the love-seat calls;
after eight, sleep falls.
(Coffee may inhibit drowsing,
but isn't positively arousing.)
Take risks? Certainly, Miss Risks,
where would you like to go?
Burning at both ends ah! and oh!
it gives a lovely light,
but will the game prove worth the candle,
let alone burnt fingers and a scandal?
Will be? Won't be? On the whole
you might find more delight
in questing for the soul –
that haggard sexy heroine,
the eternal feminine.

Where can it be?

The seat of the soul. Hardly in the purlieus given over to pleasure. Possibilities were struck off one by one: the nose, blocked with grime, prone to offend its owner; the eye, ever looking for offence; the all-too-busy hand. (How can it be, as a modern sage declares, that the body is the best picture of the soul?) Until only the zones of pain remained. Should one suffer solely from gout, would the soul therefore reside in the big toe? Well then, the breast, or the heart? Once highly desirable neighbourhoods but now overrun by lawless pangs and migrant uneases. Or the head, the capital itself? Smoggy, littered with grubby worries, addicted to aspirin and other drugs. Or, as some holy man surmised, the pineal gland? A 'pea-sized organ in the brain'. Roominess was immaterial, but even so, not a very grand address. If no suitable seat could be located – would this suggest there was no soul?

Secret drawer

The secret drawer locked against burglary,
What did it conceal? Nothing so innocent
As a prescription for glasses, a rosary.
No, but a bundle of letters his agent
Had written him over the years, doggedly
Urging him, with little sign of resentment,
To tackle something major – a biography.

Instruments to plague

As time went by, they found you out,
In failing sight, strange warts and gout –

He traced each single dolour to its dim
Unvirtuous place, now costing him . . .

Though still aware of those, the utterly
Innocent, struck down already, finally.

That there was justice in the world he saw;
And of injustice rather more.

The evil days

While the years draw nigh when the clattering typewriter is a burden, likewise a parcel of books from the postman, and he shall say, I have no pleasure in them; for much study is a weariness of the flesh;

Also when the cistern shall break, and the overflow be loosed like a fountain; when the lights are darkened, and the windows need cleaning,

And the keeper of the house shall tremble at rate bills, and be afraid of prices which are high; and almonds are too much for the grinders, and beers shall be out of the way;

Yet desire may not utterly fail, and he shall rise up at the sight of a bird, when the singsong girls are brought low;

In the day when he seeks out acceptable words; when the editors are broken at their desks, and the sound of the publishers shall cease because they are few.

Then shall the dust return to the earth as it was, and the spirit also, whether it be good or evil, shall look for its place.

Waiting to be served

The boss can't spit in every single dish,
Or leave the bones in every fillet of fish
Or (it would cost the earth in bananas)
A banana skin for each of his customers? –

Except he is weary of numbering
The fall of sparrows and farthings.
We shall all of us get the same dessert.
No more dying alone, or lying in state –
Form up in queues and wait.

*

Don't play with me, Lord. I don't want
To be understood. Nor to hear about
Extenuating circumstances.

337

Do we all get in by the back door, then?
Please – it's neither the time nor the place
For theology. And I never asked to be saved
By others dying on a battlefield or cross.

Better to be punished in hell than patronized
In heaven. I can understand your critics.
Rather that other court of justice,
Handing down plain verdicts –

Where sin can enter by the big front door
And feel at home.
We want to feel at home, Lord.

(God as a spiteful chef? As a benign bumbler? Some people are hard to
please! It seems that what used to be known as 'religious mania' is still
found in a warped form among the irreligious.)

Last things

You need to be a Catholic to stand that sacrament
Going on around you. Otherwise, passionless faces
Of nurses, and someone you vaguely know, saying
The worst is over and you'll soon feel much better.
Probably you will. It can't hurt to leave with the
Sound of truth in your ears. Unless you guess at
Rewards or reprisals ahead, and doubt that better's
The word for you. In which case the last rites are
Not so bad. (Extreme unction has helped you before.)
Nor, come to that, is the needle, that blunt visa.
At least go in peace, since you must go, in peace.

Peace, here, being primarily a physical affair. This is the body's hour.
The soul, wherever situated at the moment, has all eternity before it.
It's the flesh that wants to be put decently to sleep. The spirit, with a
care for residual dignity, will not disapprove. May even concur. They
were closer, over the years, than was often supposed.

Ever-rolling

Though much is uttered at many assemblies
We fail to hear the holy voice of Clio.

History perhaps has ceased to make herself;
More than enough, she thinks, to last for ever.

Like brilliant fog, the dead, the famous ones,
Eat up our oxygen, and nod politely –

At every man a king, king for a day, then
Lost in a common last unfinished phrase . . .

Those quivering mounds of flesh and tatters –
Matter for some new Gibbon or Macaulay?

A score or more of leading international poets
Mount a platform. The earth declines to move.

Some thought they saw the shade of Thomas Mann
Linger by a news-stand, a pained look on its face:

'After me a deluge. Merely a flood.'
Perturbed of course, and yet a trifle smug.

While ever-rolling streets bear shades and
Solids, agents, patients, all away,

And still we sit in chains, where film
Unwinds its strings of undistinguished horrors.

If all the innocents, the slaughtered commons,
Swollen children, if only they could rise

In one great cloud to heaven, at least
The guiltless, could go and leave us free,

Leave us a yet unwritten page. A virgin year,
And this tired soil lie fallow. What might follow?

We thought we prized the past, its noble gifts,
White elephants that ate us out of heart and home.

If on the moon mankind could lose its memories;
On some fiery star our brains be wiped quite clean.

Until which time we make our unfresh starts
And share our instant chronicles. It's your turn now.

POEMS 1987

God Creating Adam

according to Blake

Either He is fatigued –
Creation takes it out of you
And this is the sixth day –
Or else He is having second thoughts.
(But God's first are always right.)

Adam has no thoughts at the moment,
Only a certain bewilderment:
It was all so sudden,
There's no substitute for a leisurely womb.

He hardly seems a sufficient abyss
To be so intensely brooded over.
How can he hope to live up to it?

The serpent is already coiled about him.
Surely a premature appearance?
(Though no surprise to the Omniscient.)

Heretics would say later, it was life
Snaking out of Adam's right foot –
As if He'd left a hole in it!
Or perhaps intended, Eve as yet undreamt of,
For a household pet. Snakes have clean habits.

Prudently God provides Adam with ribs in plenty:
No need to start from scratch next time.
'I must create a system,' He is thinking,
'And leave the rest to them . . .'
The snake is whispering about a thing called 'sex',
Though what it is he doesn't know from Adam.

Can that be meant to be a sunset –
Louring and chaotic?
It doesn't look natural,
Not exactly the shepherd's delight.

Next there would have to be a Shepherd.
That's the trouble with creating,
There's no end to it.

The Monster's Request

And so you discovered the secret of generation
At last. Though not of generosity.
I saw and heard and smelt,
For you had given me eyes and ears and nostrils,
Of a sort.
And later, with the help of a mirror,
I saw myself.
I have my feelings. They were hurt.
You could have taken a little more trouble,
Considering.
Considering that you managed to bestow a soul on me,
Or what, having subsequently learnt a number of words,
I suppose is a soul,
You could have put the bones in their proper places
(You could have robbed a better class of grave)
And moulded the flesh less casually.

But no, you say, you're a scientist,
Not one of those arty types, not an aesthete,
Not concerned with flimflam.
Your feet are the right way round, you tell me,
Count your fingers, and your blessings.

Everything is relative, you told me,
But I have no relatives,
Genesis devoid of genealogy.
Plainly you cannot be my father,
More plainly not my mother.
Hence, in all modesty, I ask for a companion.
Not a belle, not like your bride-to-be,
Or even your late lamented housemaid.

But at least a creature of the other sex,
Since sex you also gave me
(No complaints on that head, but who can see it?) –
With similar imperfections, and a few extra.
Then, since everything is relative,
I'll be quite a beau.
So – for I know the perversity of your nature –
No flimflam. Just the essentials, doctor.

An Old Story

There were to be a thousand of them, Buddhas or something akin,
And every one of them different. He had finished 999,
And they were all and each, in some particular, different.
He had worked as fast as he could, but time was running out.

Now they came puffing up the hillside, abbots and priests,
Provincial governors and the usual lackeys; also (he feared)
An executioner from the Board of Punishments, severe of mien.
He strove to imagine the thousandth and its peculiar uniqueness.
They were just round the corner, the tally-man was shouting
'990 . . . 995 . . . 998 . . .' The executioner pursed his lips,
While the priests and the governors shaped the beginnings of a
Discreet smile. The Emperor was a stickler for punctuality.

The niche at least was ready. He jumped into it, cocking his head
And raising his hands in what he hoped was a brand-new position –
With luck they wouldn't notice that the thousandth was alive . . .
It wasn't. Some benign and witty divinity turned him into stone.
The sculptor was sculpture, the master his own masterpiece.

Much later came the new, the young, the red-blooded, licensed
To renounce the past and all its works. Ancient philosophers
Had been tried and condemned *in absentia*. Recent poets
Were sent to clean out latrines. The watchword was 'Realism!'
And now came the turn of the thousand Buddhas or whatever –
Equally unreal and irrelevant – they were. Crash, crash went
The sledge-hammers, demolishing the minute particulars and
Reducing uniquenesses. There was much to be done and undone.
They hastened to the thousandth. Crash went the hammers.

See, he has lost his head. But otherwise he is all there.
Which can hardly be said of some among us standing here,
Encouraged to gawk, now that policies have been slightly reversed
And the past is permitted. And we mutter ruefully an apt
And innocuous proverb: 'Easier to pull down than build up',
Cocking our heads, unsure what to do with our hands.

Quotations

Magazines have names like *February, November, December*,
And often die within a month.
Only the *People's Daily* expects a longer lease.
'Before I was "liberated" ten wasted years had passed.'
Or, 'I was silenced for twenty-one years.'
How much would Keats have managed to write then?
Ready to die young of consumption, Qu was executed by a firing
     squad.
'The readers I once had, they believed I was long dead.'

'To write poems on a girl's leg is romanticism; to ban poems
     on a girl's leg is classicism.'
Realism is to write about blue trousers.
Ding Ling was ordered to labour in the north:
Will her solicitude for sick chickens redeem her?
Rectification ought to have put things right.
Yet Confucius is again in disgrace,
Tagore's teaching is 'morphine and toddy'.

So the inverted commas come and go –
'Traitor', 'literary despot', 'class enemy' . . .
Happy the painter who stuck to the water-buffalo,
A static symbol, 'it lives and dies for the people'.
If once and for all we knew the meaning of 'revolution',
If only 'reaction' were cut and dried,
We could write a simple history of the people.

Mixed Proverbs

Outside the restaurant, garish but inside delicious no doubt,
A most famous of Guangzhou, if not of Canton as once,
Excited citizens are stringing up garlands of firecrackers.
What they are arguing about is beyond the visitor's knowing –
When he offers his lighter they absently wave it aside.
It may be a question of how best to arrange them, or whether
Such acts are utterly licit. *The people fear the officials,*
Thinks the foreign devil, who has heard of this local saying.
For how to tell what the party line currently is on fireworks,
Their potency and periodicity, their season and station?
He remembers his childhood, and a miserable sprinkling of them,
They didn't have money to burn, and the sense of dull guilt
When a rocket winked out or a Catherine wheel stopped dead.
Come on, he wants to cry, let's have them, a good half-hour
Of continuous din, driving out evil spirits (should any remain),
And damn the expense! But the bickering goes on, matches are
Struck, then pinched out, fingers point upwards and downwards.

His official hosts, at the door, solicit him nervously, please to
Enter, the feast is afoot. *The officials fear the foreign devils.*
What the eye never saw, the heart never grieved over; he can't
Pretend he ever yearned for stewed turtle or sea cucumber.
But he thinks of the engine his father once bought for him,
And even a tender. And somehow a section of rail was wanting,
The circle incomplete: when they forced the pieces together
The engine keeled over. The open floor was the only place for it,
Unauthentic and pitiful, as it banged into chair-legs and feet.
No matter, he's off to Shanghai tomorrow. It's a big country,
It can't be short of railway track. Sometimes the years supply
Our childhood lacks. All things come to them that wait.
But he must make sure to seem small and humble, and kowtow
To the woman who keeps the key to the WC.
The foreign devils fear the people. Now the circle's perfect.

Fitting the Crime

Wedekind is incarcerated in a fortress
For committing a satire against the Kaiser.
It's really quite a fine and private lodging!
No banging of spoons and platters in a fastness,
Though meals are sometimes late arriving.
They serve him generously with pens and paper.
Sentenced to six months of hard writing . . .
Then he'll walk out, in his bulging valise
A nightshirt, a dayshirt, and a manuscript.
Someone will publish it, others will read it.
To live at the *fin du siècle* was all but bliss!

Writer's Tinnitus

So it's come to this! Not only when you've
Just brought out a book, and not quite
Faintly praising: more like a hissing,
Not especially faint, and minus respite.
   In sweeter terms, it's what you still remember
Fondly: the long continuous chirring
Of cicadas, through the night, in one or other
Long-left place – that tropic velvet,
That fecund heat – making a second, more stirring
Sort of silence. But what enhances darkness
By day grows merely stupid.
   Or you might say, a distant soughing,
A sea whose waves unbrokenly are breaking;
Or some well-mannered but unfaltering wind
Through brittle grass; or gas escaping.
   Or tittle-tattle's buzz: a new biography
In the offing, rending of veils and shrouds?
(For truth is great and shall prevail,
If stiffened with pornography.)
   Or else, at speed, a tape of all the words
That ever were, stripped of all their affect.
Tools of the trade reduced to rusty susurration,
Ancient arid bookstacks calcining unchecked.

But this is mere imagination
(A presence which itself is hissing, darkly
Whispering in the wings, without remission,
And lucidly, alas) –
And you no better than the class of ass
Who bores his doctor with prodigious symptoms!
    Still, something's cooking. A pullulation
Of micro-organisms, non-readers all the lot,
All ranters? Then it's when the ululation
Ceases that you start to worry. Or you stop.

The Decline of Science Fiction

started when in place of plots and brain
sex raised its oldhat head over heels
magnificent spacemen in their flying suits
ripped off and up to same old tricks
vaster than empires and more slow
in foreign parts, black holes and asteroids
whereon effects of zero-gravity
leading to levity, the big bang theory

while alien life forms watched
and wondered what was taking shape
(like kiddies: is daddy killing mummy?
or grandads: tut, how culture-shocking!)
the hippomorphs were frightened
small chance of interbreeding though
forbidden by galactic councils or
savages finding nobler things to do
and crudest robots never screw

they have their own techniques
by clone alone or else their bodies think
or wholly merge like Milton's angels
Hindrance find none of membrane, joint or limb
Who are the monsters now? here was the decline
and fall of sf, science and fiction
caused again by sex, all art's ruination
curse of the earthmen.

Poetry Readings

These are very meaningful, for the public can learn
How the poet imagines the words are pronounced and
Which are the ones that actually signify something.
Moreover there is always a chance that while thinking
Or drinking he (or less commonly she) will fall off
The platform, or dissolve into self-induced sobbing.
Or at least flirt flauntingly with an attendant mistress
Who of course is beautiful, attractive and farouche
(In the words of Louis Simpson); and probably under age.

Unless, as may happen, the public is utterly ignorant
Of this important and rewarding event, since
The organizers have omitted to make it public.
In the grip of passion, the secretary and the treasurer
Have fled clandestinely to the nearest watering-place.
Or the committee has removed to some distant hostelry,
Leaving a notice nailed to the door: 'Mr X is sick',
And applying the travelling expenses to improper uses.
Or quite simply and innocently the date was misprinted.

No skin off your nose if (as old records have it)
The Latter-day Saints are rising in an adjacent chamber,
Or a celebrity from Assisi by the name of Francis
Is on stage round the corner with performing animals,
Or the Old Fire Station Arts Centre has gone up in flames.
Such misfortunes can be borne with fair equanimity.
More likely that while you were making your journey
A blizzard was predicted, and the prudent stay indoors,
Or among 'the many keen admirers of yours hereabouts'
Only two are about, two of the less admiring.
When Mr Y was there, they tell you, the place was packed,
Loudspeakers were erected in the street outside.

Yet always there is someone present, one who deters you
From calling it a day and taking the night train back.
Not the cultivated couple who offered accommodation,
But a woman proclaiming 'I am the Laureate of Leigh-on-Sea',
Or a man who has always loved your poem called 'Crow'
Or a book you once wrote whose title, he remembers,
Featured ambiguities in a number of types.

For there is no case on record when no one at all turned up.
Not as yet.

Joys of Writing

Memo as requested on incidence of suttee in district for which responsible under Crown. Deplorably rife and impossible stamp out being commonly approved, as witness Sanskrit expression: 'keeping company'. Yet fear some have arms and legs bound together and are carried, not through grief or debility, to pyre. I say, this is a – a story! How well I am telling it.

*

Regret to inform died in action. Gallant soldier, sad loss to fellow officers and men. Do next of kin need to know circs? Storming machine-gun singlehanded. Trapped in burning tank. Hanging on wire under heavy fire. (Query rhyme.) Go easy on details, e.g. head or legs blown off. Imagination kinder. Peaceful smile on youthful face . . . Thank you Imagination, and my old literature teacher.

*

Dear Diary, these are the last words I shall ever confide to you, my life lies in ruins. But not half as ruined as that silly boy besotted by serpent's smiles. Ruined already she was, and no chicken, her and her padded bosom and fatty thighs. He'll find out soon enough and come running back. Ha! I feel better already, dear Diary. More tomorrow.

*

Have never really enjoyed death sentences, but somebody has to write them . . . Sort of fellow one admires, stuck stoutly to his beliefs. Wrong ones. Mention with regret reasons unmentionable for security reasons, greater good greater number, sleep safer in beds, etc. Death sentences are usually short, but this deserves a paragraph – who would have thought, an inspiring theme! – and me to write it.

*

Put off long enough, must write and tell her once and for all, all is over. After so long – will she take it amiss? Excitable woman prone to impulse. Suppose that was her charm. Ah it's coming – an intimate (but not too) memory, a lightly concealed compliment, some rueful but wise reflections . . . Must make a copy to keep.

*

Depressing task, composing obituary for decent quite useful fellow. Died prematurely. No staying power. Humble beginnings. Educated Eton and King's, was it? Universally loved. Small world. So hard to express common notions in personal style! Perhaps choice quote from Horace? *Vitae summa brevis* . . . Excellent, shall put in own words, no one will know.

*

Leave it to unlettered lawyer? Not likely. Witty and weighty shall be my ultimate utterances, pages for posterity, beauty's bequest. Also chance to say what think of nancy nephew and noisome nieces. Last words should be limpid and wondrous, a will to power, a testament to talent. Writing – a life-enhancing joy for ever. My word, the Word is indeed the Deed! With which I commend my soul to the biographers.

Abbey Going
(i.m. Philip Larkin)

Poets' Corner is somewhere to stand in,
In disgrace.
Giving the devil his due,
Your intellectual integrity – says the Sub-Dean
Up there at the holy end – denied you
The faith that could alone efface
Your lifelong fear of dying . . .
Some call it honest doubt.
Complaining we come in, and others complain
When we go out.

And then the psalm. 'I will take heed to my ways,
That I offend not in my tongue.
I will keep my mouth as it were with a bridle . . .'
You didn't publish very much. '. . . kept silence, yea,
But it was pain and grief to me.' (Also to us.)
But nor were you notably idle.
Followed by the Laureate reading the lesson
From the famous Ecclesiasticus,
And then – to the relief of many – Sidney Bechet.

As a send-off, somewhat curious!
They were playing at home, you were away.
How firm and frank the Church can be,
Inside a church.
Who is the less deceived is not for us to say.
You didn't especially love the garish day,
And now 'the night is gone'.
What it adds up to is, you couldn't lose.
You might as well have kept those bike-clips on.

Happy Ending

Here is an ageing author, looking for his book,
His ancient book, perhaps his one and only.
Where can they be, we wonder, those pristine copies
Once duly supplied by some quondam publisher.
Were they burned, did they rot, were they stolen?
Did he give them (it is easy to give them) away?

Mile upon mile he trudges, to no avail,
Calling at shop after shop, sellers of books
More thick on the ground than you ever imagined.
Not one has his book, nor has anyone heard of it.
An item, it must be, of minority interest,
Or he suffers from delayed delusions of grandeur.

Then he picks up the phone. And in a mere tick
A bookshop proudly professes, it possesses a copy
Of the volume in question, long lost, long sought for!
And is prepared to dispose of it, for a price
Not disclosed to us. But who cares about money?
His wrinkled old face is wreathed in new smiles.

This frankly is a commercial for the Yellow Pages,
British Telecom's gift to us, costing us nothing.
Yet we are pleased by a story with a happy ending,
Glad when old authors recover their youthful rages.

Event

Here is a picture of the helicopter
From which the pictures are being taken
This is a shot of another helicopter
From which the shot of the helicopter
From which the pictures were taken
Is taken.
This is a picture of the chairman
Of the sponsoring body (whose product
Is not to be pictured of course)
But here we can see the officials
The managers, the coaches, the doctors
And that was a glimpse of the vehicle
From which those pictures were shot

Next a brief view of a typical viewer
Who in general possesses a licence
(Don't look, he is adjusting his set)
And now we go back to the pictures of
To the pictures . . . the pictures
You have all been waiting to see
But first a look at your commentator
Who deserves to be shot . . .

Great Temptations

Maids and matrons numbering some fifteen hundred
Were waiting in their aprons to be married to Jesus.
But not a whit was Mistress Southcott flustered.

Not a wit, Satan reminded himself. It was useless
To insinuate that the Son of God was no polygamist.
'O come, Lord,' they were singing in sisterly chorus,
'Come quickly!' Was this what they meant by Methodist?

Why was he fated to strive with this female, he wondered.
Simply so she could boast of repelling his advances? –
Like the blowing of rams' horns, or so her lot reckoned,
Would repel the Frenchies. If only she'd try her chances
With Boney! But worldly emperors weren't up to scratch.

The Lord, he proposed, might appreciate a better class
Of conversation, supposing he had time for it. *Spirit
Shall speak unto spirit*, she said. But if it came to pass
That the match was postponed, then how would the ladies
While the long hours away, asked the Father of Fibs –
In artistic pursuits, so proper to prospective brides?
They would talk of babies, they would sew and make cribs.

Not up to scratch, said the Fiend archly. Listen, Joanna,
I can make you an eminent stylist, your books best sellers.
*Like Mr Macpherson, like Mr Sterne, like Pamela and Shamela?
No*, she cried. *Style is a snare, meaning is what matters!*

He had failed again. Again she would wallow in sanctity.
All he could do was slip a lottery ticket among her relics.
Why did the crazy prosper? Why were the stupid so happy?
(Those to come would credit her existence but not Old Nick's.)

Agape

motion: That lightning did strike York Minster, destroying by fire the South Transept.

argument: That through divine intercession the firemen were enabled to save all but the South Transept.

dissent: That God was displeased concerning the appointment of a new Bishop of Durham.

rebuttal: That the Lord does not conduct himself in human fashion; also, as the Archbishop of York confirms, we do not live in biblical times.

interruption: And why not, may I ask?

question: Why did God not strike Durham Cathedral? Is it suggested that His aim is uncertain?

answer: In His mysterious way He reveals that He moves in a mysterious way.

question: Though less mysterious, would it not have been more to the point to strike down Arthur Scargill?

rebuke: The Archbishop of Canterbury does not care for talk of divine intervention unless properly vouched for.

interjection: To strike Arthur Scargill with lightning would be justice; to spare the larger part of him, coals of fire.

shout: If I was Maggie Thatcher I wouldn't stand under no tree!

response: The Church of England is like the British Royal Family, above and beyond politics.

objection: Why bring the Church of England into it? Is God a member of this body?

proposition: Might it not be said that God gave freedom of choice to lightning as to Adam, Eve *et seq.*?

shout: Did He give the Bishop of Durham freedom of belief?

piteous cry: Why all this ridiculous fuss about an old pile of stone and timber? What about –

the chair: At this point I really must intervene – *[drowned by noise of thunder]*

Psalm for Supersunday

SUNDAY is for family and fellowship, and the gates of the Supermarket are open to the faithful, to families and fellows.

For this is the House of Supergod, prepared by the hands of the Supersons of men, mindful of families and fellows and their faithful credit cards.

There on the right you shall find bread, white and brown, sliced and unsliced; and on the left new wine in new bottles, made to make men glad. Vinegar is displayed elsewhere and, in Toiletries, sponges.

This was somebody's flesh and blood, they say, speaking metaphorically. The Supermarket, as likewise the lesser clergy, has set its face against metaphors, save in promotional literature. The beef is immaculately presented, not conceived; the lamb will never rise again.

Sunday is for the family and for togetherness; for babies and fresh farm eggs in gleaming chariots, with frozen fishes and hot spices, Golden Wonder and milk and honey, hyssop and Fairy Snow and ointments for the feet.

Yet those who steal on the sabbath shall be punished twofold; but blessed are the meek, for they shall inherit Special Offers. Families that pay together stay together.

The Supertemple is forever full: voices are raised in rejoicing and praise, glorious things are spoken of it.

Bow your heads as you leave, past the tabernacle of cash registers, and the wise and calculating virgins. Hearken to the sound of bells!

Prayer

O Cindy
You who are always well-groomed and cheerful
As we should be
Who tie your hair back before going to ballet school
As we should do
Who take good care of your costly vestments
As we should take
Who is put to bed and made to get up
Who speak in silent parables
Concerning charm and deportment and a suitable marriage
    to a tennis star and yachts and a rich social life
Who grow old like us
Yet unlike us remain for ever young
Whose hair is torn out at times
Whose arms are broken
Whose legs are forced apart
Who take away the sin of the world
To whom a raggedy doll called Barabbas is preferred
You who are scourged
And given vinegar to drink from a jar of pickled onions
Who seem to say, Why have you forsaken me?
Who like us may rise or may not from the dead
    in a long white garment
You after whom the first day of the week is almost named
Into your hands we commend ourselves.

Bay-Tree and Fool

The green bay-tree is flourishing.
The light from its leaves drips lividly
Down on a fool, who muses on fairness.
A fool I was born, he says modestly.

The green bay-tree thinks, I am filthy
With foliage, my sap runneth over.
At times I might wish to loiter palely,
But money just grows on me, so to speak.

355

A born fool, says the fool under the tree,
I shall die one too. And at the last,
If I am lucky, I shall have nothing
But peace. Meanwhile I dream of equity.

In self-knowledge we tie: but he talks and
Is heard; I think and am not, thinks the tree.
Indeed we shall all have peace at the last.
For the present I flourish disgustingly.

Reaching after Reason

They lack the strength to bury the dead,
Even the ones who need so little space.
Too weak to shiver, children die of cold.
God moves, unmoved, in mysterious ways.
These have not sown but still they reap.
At night when at last he falls on his bed
The doctor reaches for Sherlock Holmes –
A mystery solved, it helps him sleep.

Murder in the Sitting-Room

As for dying, our television will do that for us.
Two children are missing. Such pretty faces, and
Pretty places where they're looked for too. Such
Bustling dogs. (A nature film?) Now a small body's
Found. (Whose is it? Oh yes, it must have been a week
Or more ago.) And so the slow excitement of the hunt,
Policemen shaking heads, speaking words of caution.
Now lacy ribbons mark a spot. A third child disappears,
Or fourth. Some other little girl, same satchel at her
Side, will walk the same way home – to jog our memory
Of what we see on television. Men rake (Chicago cops?)
Through rubbish dumps. Who watches elsewhere avidly?
Whose parents weep for pretty someone else? No doubt
The Recording Angel knows, him with his bulky book.

The Lady Orders her Funeral

The robe – of this material,
Wouldn't you agree?
And here we have the measurements.
As for what our clever friends would call
The context not the contents –
It needn't be too roomy.
I'm very slim, you see.
Oh I was always slender,
That's nothing new.
Pretty as a picture –
Prettier than many,
My painter friends would say.
So, not much colour,
But that's a job that I can do.
With my peculiar profile
Though, shouldn't it rather be
Embalming in the old Egyptian way?
Oh my dear, do try to smile!
Regarding the procedure –
Words are inclined to get under the skin;
Music's best for holding things together,
A sort of safety pin.
As for the rest, no need to bother,
It's just routine.

Failing

Almost stone deaf she is now, unless
Preferring to hear nothing further
Of our sorry world, all victims or crooks.
Tired, I guess, of people's cleverness,
As mine must once have fretted her.
She has had her share of cares.
She listens, she tells me, to Talking Books.
I look at her tapes: *The Sea, The Sea* –
Oh mother, I was never clever like that!
I recall her saying: 'That Pam Ayres,
Isn't she doing well with her poetry!'
(Something for others to wonder at.)
Now her eyes are failing, can't make out TV.
She's had enough, of woes, joys, scares,
And for today, at least, of me.

Monkey Puzzles

Monkey recurs

He has been bad again.
Vanity, anger, drunkenness, tattle,
Avarice – he coveted the old age pension –
And concupiscence. In a nutshell.
Also a brief bout of sloth
Wherein he committed very little.
Bemused by a high-slit skirt
He travelled three stations without a ticket;
Or was it a low cleavage?
His memory is bad as well.
He leafed through dirty books in a shop,
And didn't buy them: meanness is wicked.
He abstained from taking life,
And instead took umbrage.
Seeking to purge and to purify
He came unstuck through attachments.
Buddha is not pleased with him.
'Egotism, envy, spleen and lust,
The iron itself creates the rust.'
Dharma may not be denied.
'Bound to the Wheel' is the Lord's sentence:
'The whole thing all over again!'
Thank you, thinks Monkey, hanging his head.
(He was good at concealing pride.)
He remembers to groan a piteous amen
As befits the justly convicted.
Nor would he wish to be thought a Nietzschean –
At least he had never envied Superman.

Hard questions

Did Eve possess a hymen or Adam a prepuce?

Did Eve menstruate in Eden?

In such ideal conditions why did she fail to conceive?

Must man fall to be fruitful?

Was Adam right to leave his father and his mother and cleave unto his wife as tacitly agreed with God?

Was it wise to want to be wise?

Why did not God do more?

What this world needs, said Monkey, *is a new theology, more relevant to this moment in time.*

Will democracy work in the end?

What is work and when is the end?

Should marriage be abolished in order to bring down the divorce rate?

Does colour count, and if so how many blacks make a white and vice versa?

Does murder matter if (a) man is an immortal soul (b) he dies before long anyway?

Is wisdom wise?

Why doesn't the UN do more?

Sacred and propane love

Monkey was on fire,
Tongues of flame came out of his extremities.
He worshipped the very back she lay on.
Two hours of scolding for ten minutes of bliss –
It was worth it.
It was evidence of a genuine relationship.
Was she married? he nerved himself to enquire.
Of course, she snorted, else she wouldn't be there,
She'd be looking for a husband.
He glowed all over. He was mad about Else.

He thought of her all the time she wasn't there.
When she was there all thought fled,
He stood on his head on the bed.
To behave shamefully it helps to be shameless.
He worshipped the knickers she never wore.
Four hours of abuse for five minutes of bliss –
That was love for you.
He was burning all over,
Including the most applicable places . . .
Countless hours of cold showers.

Storm in teacup

'And there they sit,' said Goethe the Great Gorilla, of tea-drinking in
England, 'all looking so comfortable, so pale, so beautiful, and so tall.'
Monkey scrounged an invitation, he being fond of comfort, also of
beauty and pallor and fine upstanding.

A lady looks embarrassed. Someone is pinching her pale, beautiful
thigh. Her spouse grows red in the face. Tea is spilt over a new frock.
No it won't stain. Oh yes it will. A dispute breaks out over milk or
lemon, insinuations of U and non-U, the past is read in tea-leaves.
The spout of the teapot is stopped up. Not so the tongues. Offence is
given and taken. The ants are in the sugar. The pussy is at the cream. A
hand is half-way up a skirt. The husband affects to drop his maid of
honour, bends down to investigate. The table overturns. The ceremony of innocence is drowned. A minor heart attack occurs, a best jug
breaks.

And there they squirm, all looking so uncomfortable, and blushing pink, unbeautiful, and feeling small. Though smart at balancing a cup and saucer with a tart in either hand, Monkey thinks the whole business overrated. The Witch's Kitchen is more his scene.

Olympics

He came in first in heat after heat
    because the heat was too much for the others

He won a medal for boxing
    since his opponent had broken a thumb at a disco

He carried the day in the relay races
    when the rest of the field fell down in a heap

He scored a triumph in the hop, skip and jump
    on account of the other competitors catching a virus

He came off best in throwing the discus
    since his rivals were disqualified for pulling faces

He romped home in the swimming without any effort
    for the other swimmers had lost their body fluids

If there were two or three events he didn't quite win
    it was simply because of his sex being wrong

He faced the cameras with a self-effacing smirk
    and turned off the telly and went up to bed.

In the shop

Sophie-Ann has movable arms and legs;
Also she drinks and wets.

So does Monkey;
Who is also rheumaticky.

A little girl once pointed at him:
'Can I take it home, mam?'

Mam winced and looked the other way.
The toys they made for children these days!

God's present

The Mystic Monkey taps his watch:
Time for a meal,
And after read the wicked city news
And change a washer on the tap.
God's present in the wash-basin –
Would you have Him dripped upon?

The Mystic drinks his coffee at the window,
Sees a pretty lady swinging down the street.
She drops a handkerchief.
His feet have borne him to the door . . .
His feet, as often, go too far.

There's much high talk of love and loving,
Of burning hearts and other parts,
The bridegroom coming and the waiting bride.
What's good enough for female saints
Can't be too quaint for him.
God lurks in a mysterious sway . . .

But then, that's just Teresa's metaphor.
All mystics have their pride,
They like to be original.
He turns to Radio 3 and Handel,
Music while he washes dishes.

The hankie blows along the street –
A screwed-up paper tissue.
No good pretending God's in that.

Grim Tale

In the good old days, when wishing still worked,
There lived a Poet who wished that everything he imagined
Would come true.
And so it did. Life became peaceful and prosperous
And kindly and nice. Too nice. It wouldn't do.
But he couldn't adjust his imagination,
So he wished for a Novelist instead.
And the Novelist wished that everything he imagined
Would come true.
And so it did. Life became brutish and base
And cruel and nasty. Too nasty. It wouldn't do.
But a new day had dawned, when wishing ceased to work
And things didn't come true,
And what you wished could only be imagined.

Jacket photo

Washed-out eyes, half-dead pan,
Floppy lip and furtive look –
Would you buy a second-hand book
From this man?

Spotting

A scholar spots a similarity of phrasing
In passages by a couple of Romanian poets,
One of whom survived a concentration camp
Which the other contrived to escape.
Who came first? Which abused the other?
Both are dead now. We shall never know
For sure. But what sharp eyes a scholar has!

Not even a Ph.D. and years of training
Will help us to spot that a child
Has a black eye, its arm has been broken,
Or its femur, or it wears a haunted look.
There are so many children around, and
The young are always breaking things.
It takes a sharp eye to know for certain.

On his rights

Monkey rights are monkey business,
Simply simian. We cook our geese
And settle our hash.
We are Primates inter pares,
Our grunts are full of grace.
We don't ape others. Don't ape us.
It's nuts to you.

Waiting for the Bureaucrats

When high appointments are made in the capital
All short lists must include
At least one woman.
All short lists must also contain
At least one person of a colour designated as coloured,
One homosexual of either sex,
One adherent of a religion not designated as established,
One unmarried mother and likewise father,
One candidate with a first-class degree
And one with no formal education,
Also an unemployed person.
There should always be a genuine desire
That these categories are fully and properly represented.

In the capital senior positions remain unfilled.
Minor officials doze at their stagnant desks
While ad hoc committees determine the composition
Of the interviewing boards.
These, it is argued, should include
At least one woman,
At least one person of a colour . . .
Also individuals with specialist knowledge
In the relevant area, for instance
Nominees of the War Office and the Peace Society,
A driver and a pedestrian,
An industrialist and an insolvent,
A physician of long standing,
An invalid of long suffering,
A Bereavement Counsellor and a Family Planner,
Members of minorities not otherwise specified,
And whenever possible a spokesperson for the silent majority.
There should always be a genuine desire
That these representatives are fully and properly categorized.

Meanwhile the outer provinces flourish.
The inhabitants are ruddy-cheeked or else begrimed,
All, in varying degrees, are coloured.
Some of them are patently women, others are men,
And the remainder elude definition.
Religion, like sex, is a private concern.
They meet in the streets and taverns and converse or not.
Once they would make rare visits to the capital,
Tremulous with resentful excitement.
Now, they say, it isn't worth the trouble of invading,
Running costs would be exorbitant.
In any case they are far too occupied,
Working and playing, obeying their genuine desires.

Undertones of Almost War

New Israel could so easily
Bomb Alexander's antique city.

At nights unnerved recruits
Shot at astonished bats.

But would it? With so many Jews
Among the populace.

What to do with them? Expulsion? Internment?
Leave them be in their well-lit apartments.

*

The police force struck for better pay,
Leaving the city to riot and looting.

In came the army to put down the rioters,
The looters, the police; also the students,

Who (better the devil you know)
Had thrown in their lot with the latter.

But the army had most of the arms,
So a scattering of looters were potted,

And a foreign professor whose glasses
Glinted nastily at his window.

It was a great war, but it didn't last.

*

First they dropped the bombs,
Then they disinfected the water supply.

The tea masters were affronted:
You couldn't perform ceremonies with chlorine.

Also they were proposing to heal
Their victims. They were inviting maidens,

The less badly scarred, to their homes.
Did they want a moral victory as well?

*

Bangkok quiet on departure. Then at Chiang Mai
The airport ringed by listless soldiers.

Revolution afoot in Bangkok! –
No congregation of more than five persons, please,
Except for educational talks to primary schools.

Back in Bangkok, and all quiet.
Coup had come, had gone.

                    *

The long uncluttered boulevards
Built by the Raj
Are a godsend –

A few guns can command them,
Wipe out the enemy
Expeditiously.

Who is the enemy?
(The Raj doesn't think of returning.)
'Our own people, of course': scornfully.

                    *

At Siem Reap the plane landed
Incontinently, oxygen masks and bags
Falling about our ricked necks.

'As if we'd been shot down,'
Muttered a bruised passenger.
It was a joke, it was peace-time.

                    *

The story in the streets:
Sukarno plans to bring us to our knees
By dropping swarms of prostitutes by parachute.

The British are standing by to help us.
One of them, in liquor, talks of tight precautions,
Stiff resistance, potent counterstrokes.

                    *

Malay policemen bring in
A batch of Chinese
Charged with breaking the curfew.

Chinese officers confer:
'Let's go arrest a few Malays.'
Race riots call for strict arithmetic.

*

The hundred flowers encouraged to blossom
Were thereafter mowed down in their thousands,
Or dispatched to waste their sweetness on latrines.
Flowers have learnt since then to blush unseen.

*

A babe is born with a bullet in its side!
Being Christian, *In hoc signo vinces*.

*

Barely room enough to land a plane,
Not tomb enough to hide the slain –
There's still the sea, that huge container.

*

Riots take place on television,
Flames in the darkness make a stirring show.
They say a number of people have died,
But in the darkness how would you know?

*

All those years, death close at hand,
At work, at play, at the bed's side –
An old watch rendered luminous
By the grace of radium.
Death's sallow grin in the shadows,
Minute-by-minute memorandum.
Keep at arm's length, Chronos,
Get thee behind me.
Come, sundial, waterclock and hourglass,
Lest our time run out untimely!

OXFORD POETS

Fleur Adcock
Yehuda Amichai
James Berry
Edward Kamau Brathwaite
Joseph Brodsky
Basil Bunting
D. J. Enright
Roy Fisher
David Gascoyne
David Harsent
Anthony Hecht
Zbigniew Herbert
Thomas Kinsella
Herbert Lomas
Medbh McGuckian
Derek Mahon

James Merrill
John Montague
Peter Porter
Craig Raine
Tom Rawling
Christopher Reid
Stephen Romer
Peter Scupham
Penelope Shuttle
Louis Simpson
Anne Stevenson
Anthony Thwaite
Charles Tomlinson
Andrei Voznesensky
Chris Wallace-Crabbe
Hugo Williams